INTRODUCTION TO SPECIAL RELATIVITY

INTRODUCTION TO SPECIAL RELATIVITY

WOLFGANG RINDLER

University of Texas at Dallas

CLARENDON PRESS · OXFORD
1982

Oxford University Press, Walton Street, Oxford OX2 6DP

London   Glasgow   New York   Toronto
Delhi   Bombay   Calcutta   Madras   Karachi
Kuala Lumpur   Singapore   Hong Kong   Tokyo
Nairobi   Dar es Salaam   Cape Town
Melbourne   Auckland

and associate companies in
Beirut   Berlin   Ibadan   Mexico City

Published in the United States
by Oxford University Press, New York

British Library Cataloguing in Publication Data

Rindler, Wolfgang
  Introduction to special relativity.
  1. Relativity (Physics)
  I. Title
  530.1'1   QC173.55

  ISBN 0–19–853181–8

Library of Congress Cataloguing in Publication Data

Rindler, Wolfgang, 1924–
  Introduction to special relativity.

  Includes index.
  1. Relativity (Physics) I. Title.
  QC173.65.R56   530.1'1   81-18748
  ISBN 0–19–853181–8 (Oxford University Press)
  ISBN 0–19–853182–6 (Oxford University Press: pbk.)
              AACR2

Photo Typeset by
Macmillan India Ltd., Bangalore.

CONTENTS

PREFACE

Apart from being a vehicle for communicating my joy in the subject, this book is intended to serve as a text for an introductory course on special relativity, which is rather more conceptually and mathematically than experimentally oriented. In this context it should be suitable from the upper undergraduate level onwards. But the book might well be used autodidactively by a somewhat more advanced reader. It assumes no prior knowledge of relativity. Thus it elaborates the underlying logic, dwells on the subtleties and apparent paradoxes, and also contains a large collection of problems which should just about cover all the basic modes of thinking and calculating in special relativity. Much emphasis has been laid on developing the student's intuition for space-time geometry and four-tensor calculus. But the approach is not so dogmatically four-dimensional that three-dimensional methods are rejected out of hand when they yield a result more directly. Such methods, too, belong to the basic arsenal even of experts.

In fact, the viewpoint in the first three chapters is purely three-dimensional. Here the reader will find a simple introduction to such topics as the relativity of simultaneity, length contraction, time dilation, the twin paradox, and the appearance of moving objects. But beginning with Chapter 4 (on spacetime) the strongest possible use is made of four-dimensional techniques. Pure tensor theory as such is relegated to an appendix, in the belief that it should really be part of a physicist's general education. Still, this appendix will serve as Chapter '$3\frac{1}{2}$' for readers unfamiliar with that theory. In Chapters 5 and 6—on mechanics and electromagnetism—a purely synthetic four-tensor approach is adopted. Not only is this simpler and more transparent than the historical approach, and a good example of four-dimensional reasoning, but it also brings the student face to face with the 'man-made' aspect of physical laws. In the last chapter (on the mechanics of continua), the synthetic approach is somewhat softened by a heuristic three-dimensional lead-in.

In the discussion of electromagnetism I have reluctantly adopted the SI units now so widely used in spite of their awkwardness for the theoretician. But I have indicated how the equations can easily be translated into their Gaussian (c.g.s.) forms in terms of which most relativists think. A commitment to follow a consistent notation (capital letters for four-dimensional and lowercase for three-dimensional tensors) resulted in some other awkwardnesses, such as **e** and **b** for the electric and magnetic field vectors and **w** for the vector potential (since **a** was already used for the acceleration). I can only hope that the reader will give these symbols a try and not automatically transcribe them.

I should perhaps say a word on the genesis of this book. It has a predecessor after which it is loosely structured, namely my *Special Relativity* (Oliver & Boyd, 1960), which went out of print in 1975. That little book seems to have won some faithful friends and there have been frequent requests for a new edition. But when I finally attempted such an edition I realized how much my ideas—and perhaps the subject itself—had changed and how impossible it was simply to revise the old text. So I found myself much more pleasantly engaged in writing a new book, this book, though a few of the old arguments and problems have been taken over and, I hope, some of the old spirit as well. There are also ties to my *Essential Relativity* (Second Edition, Springer-Verlag, 1977). In a number of contexts I became uneasily aware that I could neither improve upon, nor omit, nor usefully paraphrase what I had already written there. So eventually (with the publisher's kind permission) I decided simply to borrow the relevant passages *verbatim*; these may account for a total of about ten pages of the present book. My conscience was somewhat eased by the fact that, in its time, *Essential Relativity* had similarly borrowed from the older *Special Relativity*.

I clearly owe much to many authors, some by now forgotten. But I would like to acknowledge the special influence on this book of W. G. Dixon, A. Papapetrou, R. Penrose, I. Robinson, D. W. Sciama, R. Sexl, J. L. Synge, H. Weyl, and N. Woodhouse. I also owe a considerable debt to my students. As just one example I like to recall the innocent class question "but what if . . . " which, many years ago, precipitated the 'length contraction paradox'—herein included.

Dallas, November 1981   *Wolfgang Rindler*

I
THE FOUNDATIONS OF SPECIAL RELATIVITY

1. Introduction

One of the greatest triumphs of Maxwell's electromagnetic theory (c. 1864) was the explanation of light as an electromagnetic wave phenomenon. But waves in what? In conformity with the mechanistic view of nature then prevailing, it seemed imperative to postulate the existence of a medium—the *ether*—which would serve as a carrier for these waves (and for electromagnetic 'stress' in general). This led to the most urgent physical problem of the time: the detection of the earth's motion through the ether.

Of the many experiments devised for this purpose, we shall mention just three. Michelson and Morley (1887, see Sec. 2), looked for a directional variation in the velocity of light on earth. Fizeau (1860), Mascart (1872), and later Lord Rayleigh (1902), looked for an expected effect of the earth's motion on the refractive index of certain dielectrics. And Trouton and Noble (1903) tried to detect an expected tendency of a charged plate condenser to face the 'ether drift'. All failed. The facile explanation that the earth might drag the ether along with it only led to other difficulties with the observed aberration of starlight, and could not resolve the problem.

In order to explain nature's apparent conspiracy to hide the ether drift, Lorentz between 1892 and 1909 developed a theory of the ether that was eventually based on two *ad hoc* hypotheses: the longitudinal contraction of rigid bodies[1] and the slowing down of clocks ('time-dilation')[2] when moving through the ether at a speed v, both by a factor $(1 - v^2/c^2)^{1/2}$, where c is the speed of light. This would so affect every apparatus designed to measure the ether drift as to neutralize all expected effects.

In 1905, in the middle of this development, Einstein proposed the *principle of relativity* which is now justly associated with his name. Actually Poincaré had discussed essentially the same principle during the previous year, but it was Einstein who first recognized its full

[1] Notes throughout the book are placed at the end of the relevant section.

significance and put it to brilliant use. In it, he elevated the complete equivalence of all inertial reference frames to the status of an axiom or principle, for which no proof or explanation is to be sought. On the contrary, *it* explains the failure of all the ether-drift expriments, much as the principle of energy conservation explains *a priori* (i.e. without the need for a detailed examination of the mechanism) the failure of all attempts to construct perpetual motion machines.

At first sight Einstein's relativity principle seems to be no more than a whole-hearted acceptance of the null results of all the ether-drift experiments. But by ceasing to look for special explanations of those results, and using them rather as the empirical evidence for a new principle of nature, Einstein had turned the tables: predictions could be made. The situation can be compared to that obtaining in astronomy at the time when Ptolemy's intricate geocentric system (corresponding to Lorentz's 'aetherocentric' theory) gave way to the ideas of Copernicus, Galileo, and Newton. In both cases the liberation from a venerable but inconvenient reference frame ushered in a revolutionary clarification of physical thought, and consequently led to the discovery of a host of new and unexpected results.

Soon a whole theory based on Einstein's principle (and on a 'second axiom' asserting the invariance of the speed of light) was in existence, and this theory is called *special relativity*. Its programme was to modify all the laws of physics, where necessary, so as to make them equally valid in all inertial frames. For Einstein's principle is really a *metaprinciple*: it puts constraints on *all* the laws of physics. The modifications suggested by the theory (especially in mechanics), though highly significant in many modern applications, have negligible effect in most classical problems, which is of course why they were not discovered earlier. However, they were not exactly needed empirically in 1905 either. This is a beautiful example of the power of pure thought to leap ahead of the empirical frontier—a feature of all good physical theories, though rarely on such a heroic scale.

Today, over seventy years later, the enormous success of special relativity theory has made it impossible to doubt the wide validity of its basic premises. It has led, among other things, to a new theory of space and time, and in particular to the relativity of simultaneity and the existence of a maximum speed for all particles and signals, to a new mechanics in which mass increases with speed, to the formula $E = mc^2$, and to de Broglie's association of waves with particles. One of the ironies of these developments is that Newton's theory, which had always been known to satisfy a relativity principle in the classical theory of space and time, now turned out to be in need of modification, whereas Maxwell's theory, with its apparent conceptual dependence on a preferred ether frame, came through with its formalism intact—in itself a powerful recommendation for special relativity.

Apart from leading to new laws, special relativity leads to a useful technique of problem-solving, namely the possibility of switching reference frames. This often simplifies a problem. For although the totality of laws is the same, the configuration of the problem may be simpler, its symmetry enhanced, its unknowns fewer, and the relevant subset of physical laws more convenient, in a judiciously chosen inertial frame.

Our main concern in this chapter will be to set Einstein's principle in its proper perspective and to derive from it the so-called Lorentz transformation equations, which are the mathematical core of the special theory of relativity. With their help we can subject the various branches of classical physics to the test of Einstein's principle, and with their help, too, find the necessary modifications where the principle is not satisfied.

[1] Proposed independently by Fitzgerald as early as 1889.
[2] Based directly on a feature of Einstein's special relativity of 1905.

2. Schematic account of the Michelson–Morley experiment

Certainly the most famous of all the experiments designed to measure the ether drift was that due to Michelson and Morley, first performed

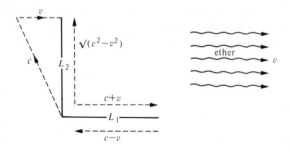

FIG. 1

in 1887 and repeated many times thereafter. Its essential principle was to split a beam of light and then to send the two half-beams along orthogonal arms of equal length, at whose ends mirrors reflected the beams back to the starting point where they were made to interfere. Then the entire apparatus was rotated in the plane of the arms. If this causes a differential change in the to-and-fro light travel times along the two arms, the interference pattern should change. Suppose originally one of the arms, marked L_1 in Fig. 1, lies in the direction of an ether drift of velocity v. Figure 1 should make it clear that the respective to-and-fro light travel times along the two arms would then be expected to be

$$T_1 = \frac{L_1}{c+v} + \frac{L_1}{c-v} = \frac{2L_1}{c(1-v^2/c^2)},$$

$$T_2 = \frac{2L_2}{(c^2-v^2)^{1/2}} = \frac{2L_2}{c(1-v^2/c^2)^{1/2}},$$

where L_1 and L_2 are the purportedly equal lengths of the two arms. Since $T_1 \neq T_2$, a rotation of the experiment through $90°$ should produce a shift in the interference fringes. None was ever observed, which seems to imply $v = 0$. Yet at some point in its orbit the earth must move through the ether with a speed of at least 18 miles per second (its orbital velocity) and this should have been easily detected by the apparatus. Of course, in Einstein's theory, this null result is expected *a priori*.

In the Lorentz theory the null result of the Michelson–Morley experiment was explained by the contraction of the arm that moves longitudinally through the ether, so that the *actual* lengths of the arms are related by $L_1 = L_2(1-v^2/c^2)^{1/2}$, which yields the observed equality $T_1 = T_2$. (It can be shown that the contraction hypothesis ensures $T_1 = T_2$ for *all* positions of the arms.) That there is also need of a second hypothesis—time dilation—in the Lorentz theory can be appreciated by considering a simple thought experiment. Suppose we could measure the original to-and-fro time T_2 directly with a clock, and suppose we could then move the arm L_2 along with the ether so that v becomes zero. Then the to-and-fro time should be $T_3 = 2L_2/c = T_2(1-v^2/c^2)^{1/2}$. But if nature's conspiracy to hide the ether is complete, we would instead measure $T_3 = T_2$. This could be accounted for by the hypothesis that a clock moving with speed v through the ether goes slow by a factor $(1-v^2/c^2)^{1/2}$. For then the *measured* time in the original position is less by that factor than the actual time T_2, and is thus equal to T_3.

As has been stressed by Sexl, modern equivalents of the Michelson—Morley experiment are being performed daily. For example, the synchrony of the seven atomic clocks around the globe that serve to define 'International Atomic Time' is continually tested by an exchange of radio signals. Any interference with these signals by an ether drift of the expected magnitude could be detected by the clocks. Needless to say, none has been detected: day or night, summer or winter, the signals always arrive with the same time delay. Again, the incredible accuracy of some modern radio navigational systems depends crucially on the directional invariance of the speed of light.

3. Inertial frames in special relativity

A frame of reference is a conventional standard of rest relative to which measurements can be made and experiments described. For example, if we choose a frame rigidly attached to the earth, the various points of the earth remain at rest in this frame while the 'fixed' stars all trace out vast circles in the course of each day; if, on the other hand, we choose a frame attached to the fixed stars then these remain at rest while points on the earth, other than those on its axis, trace out approximate circles in the course of each day, and the earth itself traces out an ellipse in the course of each year; and so on. Among all possible frames there is one class which plays a special role in classical mechanics, namely the class of *inertial frames*. These frames play an even more fundamental role in the special theory of relativity and we shall therefore define and discuss them carefully. *An inertial frame is one in which spatial relations, as determined by rigid scales at rest in the frame, are Euclidean and in which there exists a universal time in terms of which free particles remain at rest or continue to move with constant speed along straight lines* (i.e. in terms of which free particles obey Newton's first law).

Free particles placed without velocity at fixed points in an inertial frame will remain at those points, by definition. We can therefore picture an inertial frame as an aggregate of actual or virtual free test-particles mutually at rest, as determined by rigid scales. The distances between these 'defining' particles satisfy the Euclidean axioms—an important stipulation in view of later developments. Straight lines in such a frame can be defined as geodesics (lines of minimum length) and free particles not belonging to the defining aggregate move along such lines. We can further picture the defining particles as carrying clocks that indicate the universal time throughout the frame.

Now let us see the relevance of this to special relativity. We shall adopt the modern view (largely due to Einstein) that a physical theory is an abstract mathematical model (much like Euclidean geometry) whose applications to the real world consist of correspondences between a subset of it and a subset of the real world. In line with this view, *special relativity is the theory of an ideal physics referred to an ideal set of infinitely extended gravity-free inertial frames*, such as we described above. Why 'gravity-free'? Classically, gravity was regarded as an overlay which did not affect the rest of physics. So it was logical for Newton to treat the frame of the fixed stars as inertial, in the sense that *but for gravity* free particles would move uniformly relative to it. But Einstein, in his *general relativity* (the details of which are beyond the scope of this book) taught us that gravity[1] is *curvature* (of space and time) and so affects *all* the rest of physics, which has no choice but to play on a stage of space and time. In E. T. Whittaker's phrase, gravity ceased to be one of the players and became part of the stage. Thus extended inertial frames cannot be realized in nature, because gravity destroys Euclidicity. But this does not affect in any way the logic of special relativity as an abstract theory (just as it does not invalidate Euclidean geometry). It merely puts limitations on its correspondence with the real world. These are spelled out by Einstein's *equivalence principle* of 1907 (on which he eventually based his general theory of relativity): *the sets of inertial frames in the real world that correspond to (portions of) the ideal set of inertial frames discussed in special relativity consist of freely falling local frames*. At any given place and time in the real world there is one such set, each member of which can be realized by an aggregate of test-particles momentarily at rest relative to each other and falling freely under gravity. Certainly in Newton's theory such a set is locally equivalent to a set of inertial frames from which gravity has been eliminated, for in a gravitational field all particles suffer the same acceleration. Most of us have at least vicariously experienced such frames: we need only recall the televized pictures of space capsules in which astronauts are weightless and, if unrestrained, move according to Newton's first law. Such capsules, then, are the primary reference frames in which the laws of special relativistic physics would be expected to apply most accurately.

In this book *all* reference frames used (unless otherwise stated) will be ideal infinitely extended gravity-free inertial frames, and all observers will be considered to use such frames ('*inertial observers*'). Sometimes the term 'inertial' may be omitted, but it will always be understood.

For our ideal frames we shall assume certain axioms. The first is that any frame in uniform (translatory) motion relative to a given inertial frame is itself inertial. This is certainly the case in Newton's theory. Conversely, a frame *not* moving uniformly relative to an inertial frame cannot be inertial, for Newton's first law would fail in it. So the set of ideal inertial frames consists of infinitely many members all moving uniformly relative to each other.

Our next axiom is that all inertial frames are spatially homogeneous and isotropic, not only in their assumed Euclidean geometry but for the performance of all physical experiments. By this we mean that the outcome of an experiment is the same whenever its initial conditions differ only by a translation (homogeneity) and rotation (isotropy) in some inertial frame. *This is a very strong assumption*, which we are encouraged to make only in view of Einstein's relativity principle. It already eliminates the possible existence of an ether drift in any inertial frame.

It may be noted that, whereas our definition of inertial frame determines the *rate* of time (as that in which free particles move uniformly), the isotropy axiom determines the clock *settings*. For suppose isotropy holds in an inertial frame referred to Cartesian coordinates x, y, z and we define a new time $t' = t + kx$ ($k = $ constant > 0). Then Newton's first law will still hold. But any given rifle will now shoot bullets faster in the negative x-direction than in the positive x-direction (i.e. with greater *coordinate* velocity).

As a final axiom we assume that inertial frames are temporally homogeneous, i.e. that identical experiments (relative to an inertial frame) performed at different times yield identical results. In particular, this implies that all methods of time keeping based on repetitive processes are equivalent, and it denies such possibilities (envisaged by E. A. Milne) as that inertial time—relative to which free particles move uniformly—falls out of step over the centuries with atomic time, e.g. that indicated by a caesium clock.

---

[1] Actually, the rate of *change* of gravity.

4.   Einstein's two axioms for special relativity

As we have seen, Einstein's reaction to the failure of all attempts to detect the ether frame was radical. He advanced the following *principle of relativity: the laws of physics are identical in all inertial frames,* or, equivalently, *the outcome of any physical experiment is the same when performed with identical initial conditions relative to any inertial frame.*

Note that this is a generalization to the whole of physics of a relativity principle long known to be satisfied by Newtonian mechanics. Such a generalization is strongly supported by the essential unity of physics. For it would be very disturbing if, for example, the electromagnetic laws governing the behaviour of matter on the atomic scale possessed different invariance properties from the mechanical laws that govern its macroscopic behaviour. Einstein also cited instances of manifest relativity from electromagnetism. For example, the current induced in a conductor by a magnet is the same whether the conductor is at rest and the magnet moving, or vice versa.

Other *a priori* arguments can be adduced to justify the adoption of Einstein's relativity principle. But in our present development it is in fact already implicit in the homogeneity and isotropy axioms for inertial frames.[1] The demonstration of this depends on the simple lemma[2] that 'between' any two inertial frames S and S′ there exists an inertial frame S″ relative to which S and S′ have equal and opposite velocities. For proof, consider a one-parameter family of inertial frames moving collinearly with S and S′, the parameter being the velocity relative to S. It is then obvious from continuity that there must be one member of this family with the required property (see Fig. 2.) Now imagine two intrinsically identical experiments E and E′ being performed in S and S′, respectively. We can transform E′, by a spatial translation and rotation and a temporal translation, in S′, into a position where it differs from E only by a 180° rotation in S″. Thus,

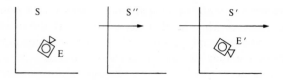

F<small>IG</small>.   2

by the assumed homogeneity and isotropy of S′ and S″, the outcome of E and E′ must be the same, which establishes Einstein's principle (in the form of our second formulation).

The acceptance of this principle—Einstein's *first axiom*—seems harmless enough until we come to his *second axiom: There exists an inertial frame in which light signals in vacuum always travel rectilinearly at constant speed c, in all directions, independently of the motion of the source.* (The value of c is 2.997 9245 ... $\times 10^8$ m s^{-1}, but $c = 3 \times 10^8$ m s^{-1} is good enough for many applications.)

By itself, this axiom is also perfectly reasonable. Even Einstein's contemporaries, familiar with Maxwell's electromagnetic theory of light, did not expect the speed of light to depend on the speed of the source, and they had empirical evidence for this axiom in their pseudo-inertial terrestrial frame of reference. In particular, the direction-independence had been very accurately tested by the Michelson–Morley experiment. But when combined with the first axiom, the second leads to the following apparently absurd state of affairs, which we shall call the *law of light propagation*:

Light signals in vacuum are propagated rectilinearly, with the same speed c, at all times, in all directions, in all inertial frames.

Thus if a light signal recedes from me and I transfer myself to ever faster-moving inertial frames in pursuit of it, I shall not alter the velocity of that light signal relative to me by one iota. This is totally irreconcilable with our classical concepts of space and time. But it was a mark of Einstein's genius to realize that those concepts were dispensable, and could be replaced by others. The final form of those others is due to the mathematician Minkowski, and consists in a certain blend of space and time into a four-dimensional 'spacetime' (1908), as we shall see in due course.

A first logical consequence of Einstein's two axioms was the elimination of the ether concept from physics. Each inertial frame now has the properties with which the ether was credited, and so it makes no sense to single out one inertial frame arbitrarily and call it the ether frame. It is true that Lorentz's theory—gentler to the classical prejudices than Einstein's, and observationally equivalent to it—kept the ether idea alive a few years longer. But soon Einstein's far more elegant and powerful ideas prevailed, and Lorentz's theory, together with the ether concept, fell into oblivion.

Finally, in spite of its historical and heuristic importance, we must de-emphasize the logical role of the law of light propagation as a pillar of special relativity. As we shall see in Section 7(ix), *a* second axiom is needed *only* to determine the value of a constant '*c*' of the dimensions of a velocity that occurs naturally in the theory. But this could come from any number of branches of physics—we need only think of the energy formula $E = mc^2$, or de Broglie's velocity relation $uv = c^2$. Special relativity would exist even if light and electromagnetism were somehow eliminated from nature. As now recognized, special relativity is primarily a new theory of space and time, and only secondarily a theory of the physics in that new space and time, with no preferred relation to any one branch.

[1] Dixon, W. G. (1978) *Special Relativity*, Cambridge University Press, p. 23.
[2] Rindler, W. (1977) *Essential Relativity* (2nd edn), Springer-Verlag, New York, p. 31.

5. Coordinates. The relativity of time

An *event* is an instantaneous point-occurrence, like the collision of two particles or the flash of a flash bulb. It will therefore be specified by *four* coordinates, one of time and three of position, e.g. (t, x, y, z). In special relativity events play a central role and we must be clear how to obtain their coordinates, at least conceptually.

The standard spatial coordinates for inertial frames are orthogonal Cartesian coordinates x, y, z. To assign these to events, the 'presiding' observer at the origin of an inertial frame need be equipped only with a standard clock (e.g. one based on the vibrations of the caesium atom), a theodolite, and a means of emitting and receiving light signals. In accordance with the law of light propagation, he can then measure the distance of any particle (at which an event may be occurring) by the radar method of bouncing a light-echo off that particle and multiplying the elapsed time by $\frac{1}{2}c$. Angle measurements with the theodolite on the returning light signal will serve to determine the relevant (x, y, z) once a set of coordinate directions has been chosen. The same signal can be used to determine the time t of the reflection event at the particle as the average of the time of emission and the time of reception.

But conceptually it is preferable to *pre*coordinatize the frame and to read off the coordinates of all events *locally*. For this purpose we imagine standard clocks placed at rest at the vertices $(m\varepsilon, n\varepsilon, p\varepsilon)$ of an arbitrarily fine lattice, where m, n, p run over the integers and ε is arbitrarily small. The spatial coordinates of these clocks can be determined once and for all by the origin-observer and then engraved upon them. To synchronize the clocks it is sufficient to emit a single light signal from the origin, say at time t_0: when this signal passes any of the lattice clocks it is set to read $t_0 + r/c$, where r is its distance from the origin. An event is now coordinatized by noting the time and space coordinats (t, x, y, z) on the clock nearest to it.

In view of the 'disparaging' remarks we have made about the law of light propagation (at the end of the last section) it will be well to point out that identical coordinates can be assigned *without* the use of light signals—though perhaps less conveniently. For example, the basic lattice could be laid out with rigid scales of equal length ε. And the vertex clocks could be synchronized by a *sound* signal from the origin if the frame were filled with still air, or by rifle bullets of known velocity shot from the origin in all directions at time t_0.

It is clear that *if* there exists a time in terms of which the physics in the frame is isotropic, our above methods have determined it, since they are based on the isotropic propagation of light, sound, rifle bullets, or whatever. Now in classical kinematics, the 'presiding observer' of a second frame S' passing through S could save himself the trouble of independent calibration by simply copying on *his* lattice clocks the (t, x, y, z) of the S-clocks with which they momentarily coincide at some instant in S. Not so in relativity: adoption of S-time in S' would lead to anisotropic physics, among other troubles. The root of this difference can be traced to the second axiom. Consider an x, t ('spacetime') diagram on which to plot events occurring on the x-axis of S (see Fig. 3). Let A, B be lattice clocks on that axis, equidistant from the origin-observer O. Now suppose the origin-observer O' of S' moves along the x-axis of S, passing O at $t = 0$, and suppose *his* lattice clocks A', B' (equidistant from O') pass A, B, respectively, at $t = 0$. Suppose further that the control light-signal of O was emitted (say at $t = t_0$) so as to reach A and B at $t = 0$. (In our diagram photon tracks have slope ± 1: we have chosen the units of time and distance so as to make $c = 1$.) Without loss of generality we can assume that O' emits *his* control signal at the time t_0' when the signal of O passes him. It is then clear from the diagram that the signal[1] from O', while reaching the coincident clocks A and A', will not reach B' until *after* it has passed B. So A' and B' must read *different* times when they pass A and B, which shows that O' *cannot* adopt the time of O! In fact, we have demonstrated the '*relativity of simultaneity*':

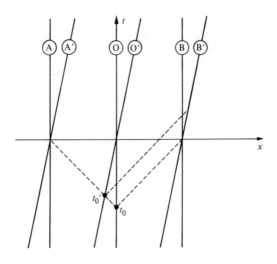

FIG. 3

whereas the events of A, A' and B, B' coinciding are simultaneous in S, they are *not* simultaneous in S'.

This detailed example has been given to prepare the reader for the surprises in store when, presently, we derive the exact relations between the coordinates of two frames.

[1] Note how the propriety of the use by O and O' of the *same* photon to synchronize their respective clocks A and A' depends crucially on the law of light propagation. No sound signal could be so used, since sound propagates isotropically only in the rest fame of its medium. But "the vacuum has no rest frame" (Lindquist).

6.   Derivation of the Lorentz transformation

In the last section we discussed methods for assigning coordinates (t, x, y, z) to events in *one* given inertial frame, based on a universal standard of time, and a universal standard of length. Such coordinates will be called *standard coordinates* for an inertial frame. We shall now consider the transformation of the coordinates of a *given event* from one frame to another. For definiteness and simplicity, let us suppose that the standard coordinates t, x, y, z in S and t', x', y', z' in S' are chosen in such a way (see Fig. 4) that: (i) S' moves in the direction of

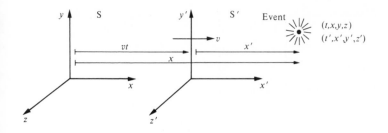

F IG. 4

the positive x-axis of S with constant velocity v; (ii) the two x-axes and their positive senses coincide; (iii) the coordinate planes $y = 0$ and $z = 0$ coincide permanently with the coordinate planes $y' = 0$ and $z' = 0$, respectively, and the coordinate plane $x' = 0$ corresponds to $x = vt$; (iv) the two spatial origins coincide when their local clocks read zero. We shall call this the *standard configuration* of two inertial frames S and S'.

In view of the already encountered non-absoluteness of time, and possibly corresponding new spatial effects in relativity, we might question whether we *can* demand (i) – (iv), especially perhaps (iii). Our procedure will therefore be to *assume* the feasibility of these conditions provisionally, and to justify this *a posteriori* by the existence of a transformation that respects *them* as well as the other axioms.

Suppose now an event has coordinates (t, x, y, z) relative to S and (t', x', y', z') relative to S'. We first briefly recall the classical (or 'common sense') relation between these two sets of coordinates, which can be just read off from Fig. 4:

$$t' = t, \quad x' = x - vt, \quad y' = y, \quad z' = z. \tag{6.1}$$

This is called the *standard Galilean transformation*. Its first member expresses the universality, or absoluteness, of time. Differentiating (6.1) with respect to $t' = t$ immediately leads to the classical velocity transformation, which relates the velocity components of a moving particle in S with those in S':

$$u'_1 = u_1 - v, \quad u'_2 = u_2, \quad u'_3 = u_3, \tag{6.2}$$

where

$$(u_1, u_2, u_3) = \left(\frac{dx}{dt}, \frac{dy}{dt}, \frac{dz}{dt}\right), \quad (u'_1, u'_2, u'_3) = \left(\frac{dx'}{dt'}, \frac{dy'}{dt'}, \frac{dz'}{dt'}\right).$$
$$(6.3)$$

Thus if a particle or signal has velocity $u'_1 = c$ along the x'-axis in S' it has velocity $u_1 = c + v$ in S: this is the 'common sense' law of velocity addition. But it is in obvious conflict with special relativity where $u'_1 = c$ must imply $u_1 = c$.

Let us see what modifications special relativity imposes on the Galilean transformation. An important first property of the transformation equations is that they must be linear [as are the Galilean equations (6.1)]. This follows from homogeneity. For consider an arbitrary clock, reading time τ, and moving uniformly (in accordance with Newton's first law) through an inertial frame S. Homogeneity requires that equal increments of τ (repetitive 'experiments' at the clock) correspond to equal increments of the coordinates t, x, y, z, which we shall temporarily denote by x_i ($i = 1, \ldots, 4$). Then $dx_i/d\tau$ = constant, and $d^2x_i/d\tau^2 = 0$. The same must be true in any other inertial frame with coordinates x'_i, say. But we have

$$\frac{dx'_i}{d\tau} = \sum \frac{\partial x'_i}{\partial x_j}\frac{dx_j}{d\tau}, \quad \frac{d^2x'_i}{d\tau^2} = \sum \frac{\partial x'_i}{\partial x_j}\frac{d^2x_j}{d\tau^2} + \sum \frac{\partial^2 x'_i}{\partial x_j \partial x_k}\frac{dx_j}{d\tau}\frac{dx_k}{d\tau}.$$

Thus for *any* free motion of such a clock the last term in the above line of equations must vanish. This can only happen if $\partial^2 x'_i/\partial x_j \partial x_k = 0$, i.e. if the transformation from the primed to the unprimed coordinates is linear. (This argument is due to Weyl.)

Another basic property of any pair S and S' of inertial frames using identical standards of length and time is that each ascribes the same velocity to the other.[1] This follows directly from the lemma of Section 4 (p. 8) asserting the existence of a frame S'' 'midway' between S and S'. For the manipulation performed in S to determine the velocity of S' can be regarded as an experiment in S''. The 180° rotation about the y''-axis in S'' of that manipulation is a possible experiment in S' for determining the velocity of S. But by the assumed isotropy of S'' the two experiments must yield the same result, which establishes our assertion ('*reciprocity*').

Now consider any event $\mathscr{P}$ and a neighbouring event $\mathscr{Q}$ (close to $\mathscr{P}$ in S *and* S') whose coordinates differ from those of $\mathscr{P}$ by dt, dx, dy, dz in S and by dt', dx', dy', dz' in S'. Suppose that at the event $\mathscr{P}$ a flash of light is emitted and that $\mathscr{Q}$ is the event of some nearby particle being illuminated by that flash. In accordance with the law of light propagation the observer in S will find that $(\mathrm{d}x^2 + \mathrm{d}y^2 + \mathrm{d}z^2)^{1/2} = c\mathrm{d}t$, or

$$c^2\mathrm{d}t^2 - \mathrm{d}x^2 - \mathrm{d}y^2 - \mathrm{d}z^2 = 0, \mathrm{d}t > 0, \tag{6.4}$$

and, similarly, the observer in S' will find that

$$c^2\mathrm{d}t'^2 - \mathrm{d}x'^2 - \mathrm{d}y'^2 - \mathrm{d}z'^2 = 0, \mathrm{d}t' > 0. \tag{6.5}$$

Conversely, any event near $\mathscr{P}$ whose coordinates satisfy *either* (6.4) *or* (6.5) is illuminated by the flash from $\mathscr{P}$ and therefore its coordinates will satisfy *both* (6.4) and (6.5). We shall deduce from this that for an *arbitrary* event $\mathscr{Q}$ near $\mathscr{P}$ the differentials satisfy the important identity[2]

$$c^2\mathrm{d}t^2 - \mathrm{d}x^2 - \mathrm{d}y^2 - \mathrm{d}z^2 = c^2\mathrm{d}t'^2 - \mathrm{d}x'^2 - \mathrm{d}y'^2 - \mathrm{d}z'^2. \tag{6.6}$$

First note that, no matter what the transformations between the coordinates may be, provided they are differentiable, the transformations between the coordinate *differentials* are linear and homogeneous. This follows from the 'chain rule' of differentiation,

$$\mathrm{d}x_i' = \sum \frac{\partial x_i'}{\partial x_j}\mathrm{d}x_j, \tag{6.7}$$

where the coefficients are just numbers at any given event. So the right member of (6.6) equals a homogeneous quadratic in $\mathrm{d}t, \mathrm{d}x, \mathrm{d}y, \mathrm{d}z$. The reader may now be tempted to shortcut the argument by asserting that two polynomials which share their zeros must be multiples of each other. But when these polynomials share only their *real* zeros, the property may not follow, as is exemplified by $\mathrm{d}x^2 + \mathrm{d}y^2$ and $\mathrm{d}x^2 + 2\mathrm{d}y^2$. So we must look more closely.

For brevity let us change notation and suppose that the polynomial

$$P = aT^2 + bX^2 + cY^2 + dZ^2 + iXT + jYT + kZT + lYZ + mXZ + nXY$$

vanishes whenever the polynomial

$$Q = T^2 - X^2 - Y^2 - Z^2$$

vanishes for real T, X, Y, Z and $T > 0$. In particular, therefore, P must vanish for $(T, X, Y, Z) = (\pm 1, 1, 0, 0)$. This yields $i = 0$ and $a + b = 0$. Similarly we find $j = k = 0$ and $a + c = 0, a + d = 0$. P

must also vanish for ($\sqrt{2}, 0, 1, 1$). This yields $l = 0$, and similarly we find $m = n = 0$. Hence $P = a(T^2 - X^2 - Y^2 - Z^2) = aQ$, as required. If we now write $c\,dt, dx, dy, dz$ for T, X, Y, Z, then Q becomes the left-hand side of (6.6), while P may be thought of as that quadratic into which the right-hand side of (6.6) transforms by virtue of (6.7). [The substituted solutions ($\pm 1, 1, 0, 0$) etc. can be multiplied by some small constant ε to respect the differential character of the variables in the above proof.] Thus at any event the following relation is satisfied by all neighbouring events:

$$c^2 dt'^2 - dx'^2 - dy'^2 - dz'^2 = a(c^2 dt^2 - dx^2 - dy^2 - dz^2),\ (6.8)$$

where a is independent of the differentials. We shall now use a device that will prove useful on other occasions too. Let us reverse the directions of the x- and z-axes in both S and S'. This does not affect (6.8), but it interchanges the roles of S and S': Fig. 4 now holds with the primed and unprimed symbols interchanged. So, by the relativity principle, we must also have

$$c^2 dt^2 - dx^2 - dy^2 - dz^2 = a(c^2 dt'^2 - dx'^2 - dy'^2 - dz'^2),$$

which then implies $a = \pm 1$. The case $a = -1$, however, càn be dismissed, since (6.8) must remain continuously true as v continuously approaches zero, i.e. as S' coincides with S. Hence (6.6) is established.

Under linear coordinate transformations

$$t' = At + Bx + Cy + Dz + E,\ (A, B, C, D, E \text{ all constant})$$

etc., the coordinate differences $\Delta t = t_2 - t_1$, etc. corresponding to two events $\mathscr{P}_1$ and $\mathscr{P}_2$, transform like the coordinates themselves *less the additive constants*, as can be seen by substituting the coordinates of $\mathscr{P}_1$ and $\mathscr{P}_2$ successively into the transformation equations and subtracting:

$$\Delta t' = A\Delta t + B\Delta x + C\Delta y + D\Delta z,$$

etc. And this is obviously also how the differentials transform:

$$dt' = A dt + B dx + C dy + D dz,$$

etc. From this it follows that the differences also satisfy (6.6):

$$c^2 \Delta t^2 - \Delta x^2 - \Delta y^2 - \Delta z^2 = c^2 \Delta t'^2 - \Delta x'^2 - \Delta y'^2 - \Delta z'^2.\ (6.9)$$

The common value of these quadratics is defined as the *squared interval* Δs^2 between the two events $\mathscr{P}_1$ and $\mathscr{P}_2$.

The crossing of origins at time zero [condition (iv)] implies that the event $(0,0,0,0)$ in S corresponds to $(0,0,0,0)$ in S'. Consequently the transformations of the coordinates are *homogeneous* (no additive constants) and therefore identical to the transformations of the differentials. It follows that the coordinates themselves satisfy a relation like (6.6):

$$c^2t^2 - x^2 - y^2 - z^2 = c^2t'^2 - x'^2 - y'^2 - z'^2. \tag{6.10}$$

By hypothesis, the coordinate planes $y = 0$ and $y' = 0$ coincide permanently, so $y = 0$ must imply $y' = 0$. But y' is of the form $At + Bx + Cy + Dz$, whence $A = B = D = 0$ and

$$y' = Cy \tag{6.11}$$

for some constant C, which may, however depend on v. Let us once again reverse the roles of S and S' by reversing the directions of their x- and z-axes. This has no effect on (6.11) but by the relativity principle we must then also have $y = Cy'$. So $C = \pm 1$ and the case $C = -1$ can again be dismissed by the consideration that S' must go into S continuously as $v \to 0$. The argument for z is similar, and thus we arrive at

$$y' = y, \quad z' = z,$$

the two 'trivial' members of the transformation. As a consequence, (6.10) reduces to

$$c^2t^2 - x^2 = c^2t'^2 - x'^2. \tag{6.12}$$

By further hypothesis, $x = vt$ must imply $x' = 0$. But x' is of the form $Et + Fx + \ldots$, so

$$x' = F(x - vt). \tag{6.13}$$

From this and (6.12) it follows that t' is of the form

$$t' = It + Jx.$$

When these expressions for x' and t' are substituted in (6.12) and the three equations that result from comparing the coefficients of x^2, xt, t^2 are solved, we find

$$F = I = \frac{1}{\pm(1 - v^2/c^2)^{1/2}}, \quad J = -\frac{v}{c^2}I,$$

where again we must choose the positive sign for the same reason as before. Thus, collecting our results, we have obtained the *standard*

Lorentz transformation equations[3]

$$t' = \gamma(t - vx/c^2), \quad x' = \gamma(x - vt), \quad y' = y, \quad z' = z, \quad (6.14)$$

where

$$\gamma = \gamma(v) = \frac{1}{(1 - v^2/c^2)^{1/2}}. \quad (6.15)$$

This is the so-called *Lorentz factor* which plays an important role in the theory.

It is now easy to verify that this transformation indeed respects all the axioms we have assumed. Moreover, it satisfies the so-called group axiom (which must therefore be already implicit in the others), which we have not yet formulated but without which the theory could not be consistent. It will be discussed in the next section.

If a law of physics is invariant under this transformation ('Lorentz-invariant'), *and* under spatial rotations, spatial translations, and time translations, then it is invariant between *any* two inertial coordinate systems and so satisfies the relativity principle. For it is easily seen that the *general* transformation between two inertial frames, whose coordinates are standard but whose configuration is not, consists of the following: (i) a space rotation and translation (to make the x-axis of S coincide with the line of motion of the S' origin); (ii) a time translation (to make the origins coincide at $t = 0$); (iii) a standard Lorentz transformation; and finally, (iv), another rotation and time translation to arrive at the coordinates of S'. The resultant transformation is called an *inhomogeneous* Lorentz transformation, or a *Poincaré transformation*. Since each link in this chain of transformations is linear, so is the resultant transformation. Poincaré transformations also evidently satisfy (6.6), since standard Lorentz transformations do, as do spatial rotations (which leave $dx^2 + dy^2 + dz^2$ invariant) and spatial and temporal translations (which leave all the differentials invariant). In fact, Poincaré transformations are precisely those that satisfy (6.6) (invariance of the differential squared interval)—and therefore also (6.9)—and contain no space or time reflections.

---

[1] For brevity of expression, we occasionally anthropomorphize frames, particles, and other inanimate objects.

[2] From this identity another standard argument for the linearity of the transformation equations can be made, see Exercise IV (1).

[3] So called because they were originally derived by Lorentz (as the transformation that formally leaves Maxwell's equations invariant).

7. Properties of the Lorentz transformation

(i) The most striking feature of the Lorentz transformation is the transformation of time, which exhibits the relativity of simultaneity: events with equal t do not necessarily correspond to events with equal t'.

(ii) Equations (6.14) are symmetric not only in y and z but also in x and ct. [The reader can verify this by writing T/c for t and T'/c for t' in (6.14) and multiplying the first equation by c.] In what follows we shall often find ct a more convenient variable than t.

(iii) The Lorentz transformation replaces the older Galilean transformation, to which it nevertheless approximates when v/c is small. This accounts for the high accuracy of Newtonian mechanics— invariant under the Galilean transformation—in describing a large domain of nature. Note also that the two transformations become the same if we let c formally tend to infinity.

(iv) For $v \neq 0$ the Lorentz factor γ is always greater than unity, though not much so when v is small. For example, as long as $v/c < 1/7$ (at which speed the earth is circled in one second), γ is less than 1.01; when $v/c = \sqrt{3}/2 = 0.866$, $\gamma = 2$; and when $v/c = 0.99 \ldots 995$ ($2n$ nines), γ is approximately 10^n. The following are useful identities satisfied by the Lorentz factor:

$$\gamma v = c(\gamma^2 - 1)^{1/2}, \ c^2 d\gamma = \gamma^3 v dv, \ d(\gamma v) = \gamma^3 dv. \tag{7.1}$$

The various proofs are left as an exercise to the reader.

(v) *Any* effect whose speed in vacuum is always the same could have been used to derive the Lorentz transformation, as light was used in our derivation. Since only one transformation can be valid, it follows that all such effects (weak gravitational waves, ESP?) must propagate at the speed of light.

(vi) Since the standard Lorentz transformation is linear and homogeneous, the coordinate differences as well as the differentials satisfy the same transformation equations as the coordinates themselves [cf. paragraph including (6.9)]:

$$\Delta t' = \gamma(\Delta t - v\Delta x/c^2), \ \ \Delta x' = \gamma(\Delta x - v\Delta t), \ \ \Delta y' = \Delta y, \ \ \Delta z' = \Delta z. \tag{7.2}$$

$$dt' = \gamma(dt - v dx/c^2), \ \ dx' = \gamma(dx - v dt), \ \ dy' = dy, \ \ dz' = dz. \tag{7.3}$$

(vii) The standard Lorentz transformation has unit determinant (as can easily be verified), and it possesses the two so-called *group properties*: *symmetry* and *transitivity*. First, direct algebraic solution of (6.14) yields the inverse transformation in the expected form:

$$t = \gamma(t' + vx'/c^2), \quad x = \gamma(x' + vt'), \quad y = y', \quad z = z'. \quad (7.4)$$

This shows that the inverse of a Lorentz transformation is another Lorentz transformation ('symmetry'), with parameter $-v$ instead of v. [Of course, (7.4) also holds in differential and Δ-form.] Next, it is found that the resultant of two Lorentz transformations with parameters v_1 and v_2, respectively, is another Lorentz transformation ('transitivity') with parameter $v = (v_1 + v_2)/(1 + v_1 v_2/c^2)$. [The direct verification of this is a little tedious; a more transparent way is indicated in Exercise I(13).] Thus the standard Lorentz transformations constitute a group. The same is then evidently true of the Poincaré transformations, as can be seen by examining their constituents—rotations, translations, and standard Lorentz transformations.

One consequence of the group properties is that if *one* inertial frame, say S, is related to all others by a Lorentz transformation, then any two, say S' and S'', are so related to each other. For, by symmetry, S' is Lorentz-related to S, which in turn is Lorentz-related to S'', so by transitivity S' is Lorentz-related to S''. And this, of course, is essential for the self-consistency of Einstein's theory. If it were not so, S would be a preferred frame.

(viii) When $v = c$, γ becomes infinite, and $v > c$ leads to imaginary values of γ. This shows that the relative velocity of two inertial frames must be less than the speed of light, since finite real coordinates in one frame must correspond to finite real coordinates in any other frame. Indeed, we can show that the speed of particles, and more generally, of all physical 'signals', is limited by c, *if* we insist on the invariance of causality. For consider any signal or process whereby an event $\mathscr{P}$ causes an event $\mathscr{Q}$ (or whereby information is sent from $\mathscr{P}$ to $\mathscr{Q}$) at 'superluminal' speed $U > c$ relative to some frame S. Choose coordinates in S so that these events both occur on the x-axis, and let their time and distance separations be $\Delta t > 0$ and $\Delta x > 0$. Then in the usual second frame S' we have, from (7.2),

$$\Delta t' = \gamma\left(\Delta t - \frac{v\Delta x}{c^2}\right) = \gamma\Delta t\left(1 - \frac{vU}{c^2}\right). \quad (7.5)$$

For $c^2/U < v < c$ we would then have $\Delta t' < 0$. Hence there would exist inertial frames in which $\mathscr{Q}$ precedes $\mathscr{P}$, i.e. in which cause and effect are reversed, or in which the signal goes backward in time. Using a 'radar' bounce off another particle with such signals, we could signal into our own past, thus foiling events which have already happened to us, and getting into deep logical trouble.[1]

On the other hand, the speed limit c *does* guarantee invariance of causality. For if two events happen on a line making an angle θ with the x-axis in S (thus not restricting their generality relative to S and S'), and are connectible by a signal with speed $u \leqslant c$ in S, we see on replacing U by $u \cos \theta$ in (7.5), that for all v between $\pm c$, Δt and $\Delta t'$ do indeed have the same sign.

We may note that, relative to any frame S, two particles or photons may have a *mutual velocity* up to $2c$. This velocity is defined as the time rate of change of the connecting vector $\mathbf{r}_2 - \mathbf{r}_1$ between the particles, which we assume to have position vectors and velocities $\mathbf{r}_1$, $\mathbf{u}_1$ and $\mathbf{r}_2$, $\mathbf{u}_2$, respectively: $(\mathrm{d}/\mathrm{d}t)(\mathbf{r}_2 - \mathbf{r}_1) = \mathbf{u}_2 - \mathbf{u}_1$, as in classical kinematics. For example, the mutual velocity of two photons travelling in opposite directions along a common line is precisely $2c$. *Arbitrarily large velocities are possible for signals that carry no information*—e.g. the sweep of a searchlight spot on high clouds, or the intersection point of a falling guillotine with the block.

One consequence of the relativistic speed limit is that 'rigid bodies' and 'incompressible fluids' have become impossible objects, even as idealizations or limits. For, by definition, they would transmit signals instantaneously.

An interesting fact, not related to the speed limit but to 'rigidity', is that a body which retains its shape in one frame may appear *deformed* in another frame, if it accelerates. As a simple example, consider a rod which, in a frame S', remains parallel to the x'-axis while moving with constant acceleration a in the y'-direction. Its equation of motion, $y' = \frac{1}{2}at'^2$, translates by the Lorentz transformation into $y = \frac{1}{2}a\gamma^2(t - vx/c^2)^2$, and so the rod has the shape of part of a parabola at each instant $t = $ constant in the usual second frame S. [For another example, see Exercise I(9).] The essential reason for this phenomenon is the relativity of simultaneity: the S-observer picks a different set of events at the rod as constituting an instantaneous view of it.

(ix) The relativistic speed limit discussed above suggests the following approach to the transformation between two inertial frames *without* appeal to the law of light propagation. Either particles can be accelerated to arbitrary speeds, or they cannot. Suppose first that they *can*. Then in S any event with $t > 0$ is causally connectible with the common origin-event $(0, 0, 0, 0)$ of S and S' (namely, by a freely falling particle). Therefore, if we assume invariance of causality, any event with $t > 0$ must have $t' > 0$, and similarly any event with $t < 0$ must have $t' < 0$. Consequently $t = 0$ must correspond to $t' = 0$, from which it follows, as in the argument for $y' = y$, that $t' = t$. In conjunction with (6.13), using group-symmetry and reciprocity, this leads uniquely to the Galilean transformation. Next suppose that particles can *not* attain arbitrary speeds. Then there must exist, mathematically speaking, a least upper bound c to particle speeds in any one inertial frame. By the relativity principle, this bound must be the same in *all* inertial frames. Moreover, the speed c—whether attained or not by any physical effect—must transform into itself. For suppose a speed c in some given direction in S corresponds to a speed $c' > c$ in S'. By continuity (surely we can assume continuity of the velocity transformation up to the maximum velocity), a speed somewhat less than c in S, and therefore a possible particle speed, would then correspond to a speed greater than c in S', which cannot be. Similarly, c cannot correspond to a lesser speed in S', and this establishes our assertion. But if c transforms into itself, each of (6.4) and (6.5) implies the other and our derivation proceeds as before, leading uniquely to the Lorentz transformation.

Thus the relativity principle by itself (together with causality invariance) necessarily implies that all inertial frames are related by Galilean transformations, or by Lorentz transformations with some definite 'c'. The role of a second axiom is now clear: it merely needs to separate these two possibilities, and—in the second case—it must also determine the value of c. In fact, the determination of c is the *only* role of the second axiom: $c = \infty$ corresponds to the Galilean transformation.

(x) What chiefly distinguishes the Lorentz transformation from the Galilean transformation is that space and time coordinates *both* transform, and, moreover, transform partly into each other: x and t get 'mixed', rather as do x and y under a rotation of axes in the Cartesian xy-plane. These aspects of the Lorentz transformation can be well illustrated on a spacetime diagram like our earlier Fig. 3, in which, however, we must necessarily ignore the two 'trivial' coordinates y and z. The diagram then directly represents events on the spatial x-axis of S.[2] Units in such diagrams are usually chosen so that

$c = 1$, and the x- and t-axes are drawn orthogonally, but this is a convention without physical significance. 'Moments' in S have the equation t = constant and correspond to horizontal lines, while the history (or '*worldline*') of each fixed point on the spatial x-axis of S corresponds to a vertical line, x = constant. Moments in S' have equation t' = constant and thus, by (6.14), $t - vx$ = constant; so in our diagram (see Fig. 5) they correspond to lines with slope v. In particular, the x'-axis ($t' = 0$) corresponds to $t = vx$. Again, worldlines of fixed points on the spatial x'-axis have equation x' = constant, and thus, by (6.14), $x - vt$ = constant. In our diagram they are lines with slope v relative to the t-axis. In particular, the t'-axis ($x' = 0$) corresponds to $x = vt$. Thus the axes of S' subtend equal angles with their counterparts in S; but whereas in rotations these angles have the same sense, in Lorentz transformations they have opposite sense.

For calibrating the primed axes, we draw the hyperbolae $x^2 - t^2 = \pm 1$. By (6.12) they coincide with $x'^2 - t'^2 = 1$, so they cut *all* the axes at the relevant unit time or unit distance from the origin. The units can then be repeated along the axes, by linearity. The diagram shows how to read off the coordinates (a', b') of a given event relative to S': we must go along lines of constant x' or t' from the event to the axes.

Spacetime diagrams like Fig. 5 are sometimes also called *Minkowski diagrams*. They can be extremely helpful and illuminating

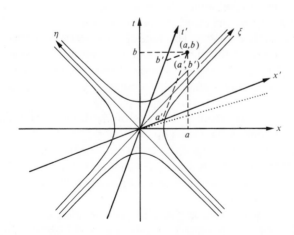

F$_{\text{IG}}$.   5

in certain types of relativistic problems. But one should beware of trying to use them for *everything*, for their utility is limited. Analytic or algebraic arguments are generally much more powerful. As a simple example of the use of such diagrams, consider a superluminal signal along the x-axis of S. In Fig. 5 this is shown as a dotted line. Since the x'- and t'-axes may subtend any angle between $0°$ and $180°$ with each other, there is a frame S' in which the signal has infinite velocity (i.e. it coincides with the x'-axis in the diagram), and others in which it moves in the opposite spatial direction, i.e. from receiver to emitter. Alternatively it can be regarded in this latter case as travelling with unchanged spatial sense into the past.

[1] Nevertheless, there have been discussions about the possible existence of 'tachyons' (from the Greek *tachys* for 'fast')—particles which in every inertial frame move at speeds greater than c. See G. Feinberg (1970) in *Scientific American* **222**(2), 68.

[2] By *spatial x-axis* we mean the set of events characterized by $y = z = 0$, not $y = z = t = 0$; it is one of the three mutually orthogonal straight lines defining the spatial reference planes of S.

Exercises I

Unless otherwise indicated, two inertial frames S and S' will always be understood to be in standard configuration. *Hint*: In working a special relativity problem, especially for the first rough time, one can often omit the cs, i.e. one can work in units in which $c = 1$. The cs, if desired, can be inserted later, either throughout the work, or directly in the answer by dimensional arguments. For example, if one established (7.1) without the cs, it would be quite obvious where to put them, since by use of cs alone the dimensions must be made to balance.

1. Establish all the properties of the Lorentz factor γ listed in (iv) of Section 7. Also draw a reasonably accurate graph of γ against v for v between $+c$ and $-c$.

2. Establish the group properties of the Lorentz and Poincaré transformations as stated in (vii) of Section 7.

3. Prove that the temporal order of two events is the same in all inertial frames if and only if they can be joined in one inertial frame by a signal travelling at or below the speed of light. Illustrate this result on a spacetime diagram.

4. If two events occur at the same point in some inertial frame S, prove that their temporal order is the same in all inertial frames, and that the least time separation is assigned to them in S.

5. If two events occur at the same time in some inertial frame S, prove that there is no limit on the time separations assigned to these events in other frames, but that their space separation varies from infinity to a minimum which is measured in S.

6. In the inertial frame S' the standard lattice clocks all emit a 'flash' at noon. Prove that in S this flash occurs on a plane orthogonal to the x-axis and travelling in the positive x-direction at speed c^2/v.

7. Prove that at any instant there is just one plane in S on which the clocks of S agree with the clocks of S', and that this plane moves with velocity $(c^2/v)(1 - 1/\gamma)$. How is this plane related to the frame S'' of Fig. 2?

8. In S' a straight rod parallel to the x'-axis moves in the y'-direction with velocity u. Show that in S the rod is inclined to the x-axis at an angle $-\tan^{-1}(\gamma uv/c^2)$. (This effect is to be expected, since we can regard the y-coordinates of the various points of the rod as clocks synchronized in S'; and synchronous clocks in one frame are not synchronous in another frame.)

9. It was pointed by M. v. Laue that a cylinder rotating uniformly about the x'-axis of S' will seem *twisted* when observed instantaneously in S, where it not only rotates but also travels forward. If the angular speed of the cylinder in S' is ω, prove that in S the twist per unit length is $\gamma \omega v/c^2$. (As in the preceding problem, the effect is to be expected, since we can regard the cylinder as composed of a stack of circular discs, each disc by its rotation serving as a clock, with arbitrary parallel radii designated as clock 'hands' in S'.)

10. Two photons travel along the x-axis of S, with a constant distance L between them. Prove that in S' the distance between these photons is $L(c+v)^{1/2}/(c-v)^{1/2}$.

11. Prove that the first two equations of the Lorentz transformation (6.14) can be written in the form

$$x' = x \cosh \phi - ct \sinh \phi, \quad ct' = -x \sinh \phi + ct \cosh \phi,$$

where $\tanh \phi = v/c$. (The reader is reminded of the relations $\cosh \phi = \cos i\phi$, $i \sinh \phi = \sin i\phi$, whereby any trigonometric identity can be converted into an identity in the hyperbolic functions.) Note that formally this is a 'rotation' in x and ict through an angle $i\phi$; as such it preserves $x^2 + (ict)^2$.

12. Prove the following additional relations between the *hyperbolic parameter* ϕ (sometimes also called the *rapidity*) defined in

Exercise 11 above and v:

$$\cosh\phi = \gamma, \quad \sinh\phi = \frac{v}{c}\gamma, \quad e^\phi = \left(\frac{c+v}{c-v}\right)^{1/2}.$$

[Recall that $\cosh\phi \pm \sinh\phi = \exp(\pm\phi)$.]

13. In the notation of Exercise 11, derive the following useful form of the Lorentz transformation:

$$ct' + x' = e^{-\phi}(ct + x), \quad ct' - x' = e^\phi(ct - x).$$

Use this to verify the group properties of the Lorentz transformation.

14. One can define alternative coordinates $\xi = ct + x$, $\eta = ct - x$, whose axes are the $\pm 45°$ lines indicated in Fig. 5. Prove that under a Lorentz transformation the directions of these axes do not change; how do their calibrations change?

15. In a frame S a slightly slanting guillotine blade in the (x, y) plane falls in the y-direction past a block level with the x-axis, in such a way that the intersection point of blade and block travels at a uniform speed in excess of c in the positive x-direction. In some frame S', in standard configuration with S, this intersection point travels in the *opposite* direction along the block (cf. Fig. 5). Now suppose in S the blade evaporates instantaneously when it passes the origin, so that a piece of paper on the block is cut on the negative x-axis only. Explain this in S'.

16. If the universe were filled with a very light waterlike fluid in which light propagated at speed $c' < c$, how would that affect special relativity? The universe *is* filled with a diffuse 'photon gas' constituting the so-called microwave background radiation (a vestige of the 'big-bang' origin of the universe). How does that differ from an ether, or from the gravitational field due to all the galaxies?

II

RELATIVISTIC KINEMATICS

8. Introduction

In the what follows we shall be concerned with the physical consequences of replacing the Galilean transformation by the Lorentz transformation. The most immediate results are to be found in kinematics, and these are already fully implicit in the Lorentz transformation itself, and involve no further modification of previously accepted physical laws. They show special relativity in its primary role as a new theory of space and time.

We shall find it convenient to distinguish between what an observer *sees* and what he can know *ex post facto*. What he actually sees or photographs at any one instant is called his *world-picture*. On an elementary level this is not a very useful concept, and quite a complicated one, since what he sees is a composite of events that occurred progressively earlier as they occurred farther and farther away. A much more useful concept is the *world-map*. As the name implies, this may be thought of as a map of the events in an observer's instantaneous space $t = t_0$: a kind of three-dimensional life-sized snapshot exposed everywhere simultaneously, or a frozen instant in the observer's spatial reference frame. When we loosely say 'the length of an object in S' or 'a snapshot taken in S' or 'a moving cylinder appears twisted', etc. we invariably think of the world-map, unless the contrary is stated explicitly. The world-map is generally what matters. These remarks are relevant in the next section, where we show that moving bodies shrink. The shrinkage refers to the world-map. How the eye actually *sees* a moving body is rather different, and in itself not very significant, except that in relativity some of the facts of vision are a little surprising, as we shall see in Section 18.

9. Length contraction

Consider two inertial frames S and S' in standard configuration. In S' let a rigid rod of length $\Delta x'$ be placed at rest along the x'-axis. We wish to find its length in S, relative to which it moves with velocity v. To measure the rod's length in any inertial frame in which it moves longitudinally, its end-points must be observed simultaneously. No such precaution is needed in its rest frame S'. Consider, therefore, two events occurring simultaneously at the extremities of the rod in S, and use (7.2) (ii). Since $\Delta t = 0$, we have $\Delta x' = \gamma \Delta x$, or, writing for Δx, $\Delta x'$ the more specific symbols L, L_0, respectively,

$$L = \frac{L_0}{\gamma} = \left(1 - \frac{v^2}{c^2}\right)^{1/2} L_0. \tag{9.1}$$

This shows, quite generally, that *the length of a body in the direction of its motion with uniform velocity v is reduced by a factor* $(1 - v^2/c^2)^{1/2}$.

Evidently the greatest length is ascribed to a uniformly moving body in its *rest frame*, i.e. the frame in which its velocity is zero. This length, L_0, is called the *rest length* or *proper length* of the body. (In general a 'proper' measure of a quantity is that taken in the relevant instantaneous rest frame.) On the other hand, in a frame in which the body moves with a velocity approaching that of light, its length approaches zero.

The statement following (9.1) is identical with that proposed by Fitzgerald and Lorentz (cf. Section 1) except that they qualified 'velocity v' by 'relative to the ether', and regarded the effect as *caused* by the ether. That the measured length of a rod is nevertheless shortened by the *same* factor in *any* frame in which its velocity is v can only be discovered by calculation in Lorentz's theory, and must appear fortuitous.

In relativity, length contraction is a kind of *velocity perspective* effect (in Weyl's phrase)—analogous to the visual foreshortening of a stationary rod that is viewed from the back rather than the top. For our world-map of the moving rod we pick what in the rod's frame are *later* events at its back than at its front, so that the back appears closer to the front. But of course nothing at all has happened to the rod itself. Just as the proper length of the stationary rod can be restored by bringing our eye into a position above it, so that of the moving rod can be restored by moving our eye along with it.

Nevertheless, relativistic length contraction is no 'illusion': it is real in every sense. For example, just as a rotated rod can be placed in the space between two parallel planes which it would not fit perpendicularly, so a moving rod can be momentarily confined in a space which it would not fit when at rest (see Section 10). Apart from the practical difficulties involved, we could verify length contraction as follows. A yard stick Y is at rest in a frame S'' relative to which two

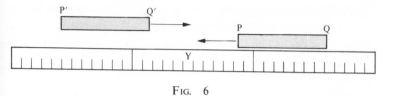

F ɪɢ.   6

identical rods PQ and P′Q′ travel at equal speeds in opposite
directions (see Fig. 6). It is not difficult to imagine a device which
leaves a mark on Y where the left end-points P and P′ meet, and also
one where Q and Q′ meet. The distance between these marks would be
found to be less than the proper length of the rods. Note that, whereas
in S″ the coincidences (P, P′) and (Q, Q′) occur simultaneously, by
symmetry, in the rest frame S of PQ the coincidence (P, P′) occurs
before (Q, Q′), and conversely in S′, the rest frame of P′Q′. This is an
obvious criterion for P′Q′ being shorter than PQ in S, and of the
converse in S′. Thus our example illustrates not only the reality and
the symmetry of length contraction, but also its relation to the
relativity of simultaneity.

Since we have no means of accelerating macroscopic bodies to
nearly the speed of light, no actual experimental verification of length
contraction has yet been attempted.[1]

---

[1] Unlike time dilation (see Section 11), length contraction is not cumulative,
so that low-speed experiments of long duration are useless: instead they
would have to be of large spatial extent.

10.   The length contraction paradox

Consider the admittedly unrealistic situation of a man carrying
horizontally a 20-foot pole and wanting to get it into a 10-foot garage.
He will run at speed $v = 0.866\,c$ to make $\gamma = 2$, so that the pole
contracts to 10 feet. It will be well to insist on having a sufficiently
massive block of concrete at the back of the garage, so that there is no
question of whether the pole finally stops in the inertial frame of the
garage, or vice versa. So the man runs with his (now contracted) pole
into the garage and a friend quickly closes the door. In principle we do
not doubt the feasibility of this experiment, i.e. the reality of length
contraction. When the pole stops in the rest frame of the garage, it is,
in fact, being 'rotated in spacetime' and will tend to assume, if it can, its
original length relative to the garage. Thus, if it survived the impact, it
must now either bend, or burst the door.

At this point a 'paradox' might occur to the reader:[1] what about the symmetry of the phenomenon? Relative to the runner, won't the garage be only 5 feet long? Yes, indeed. Then how can the 20-foot pole get into the 5-foot garage? Very well, let us consider what happens in the rest frame of the pole. The open garage now comes towards the stationary pole. Because of the concrete block, it keeps on going even after the impact, taking the front end of the pole with it (see Fig. 7). But the back end of the pole is still at rest: it cannot yet 'know' that the front end has been struck, because of the finite speed of propagation of *all* signals. Even if the 'signal' (in this case the elastic shock wave) travels along the pole with the speed of light, that signal has 20 feet to travel against the garage front's 15 feet, before reaching the back end of the pole. This race would be a dead heat if v were 0.75 c. But v is 0.866 c! So the pole *more* than just gets in. (It could even get into a garage whose length is as little as 5.4 feet at rest and thus 2.7 feet in motion: the garage front would then have to travel 17.3 feet against the shock wave's 20 feet, requiring speeds in the ratio 17.3 to 20, i.e. 0.865 to 1 for a dead heat.)

There is one important moral to this story: whatever result we get by correct reasoning in any one frame, must be true; in particular, it must be true when viewed from any other frame. As long as the set of physical laws we are using is self-consistent and Lorentz-invariant, there *must* be an explanation of the result in every other frame, although it may be quite a different explanation from that in the first frame. For example, as we shall see, the magnetic force experienced by an electron traversing the field of a permanent magnet is felt as a purely electric force in the rest frame of the electron.

[1] It is perhaps surprising that no such paradox seems to have been encountered before 1960. See: Rindler, W. (1960) *Special Relativity*, Oliver and Boyd, Edinburgh p. 37; see also Rindler, W. (1961) *American Journal of Physics*, **29**, 365.

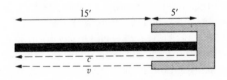

Fig. 7

11. Time dilation

Let us again consider two inertial frames S and S' in standard configuration. Let a standard clock be fixed in S' and consider the two events at that clock when it indicates times t'_1 and t'_2, differing by $\Delta t'$. We enquire what time interval Δt is ascribed to these events in S. From the Δ-form of (7.4) (i) we see at once, since $\Delta x' = 0$, that $\Delta t = \gamma \Delta t'$, or, replacing Δt and $\Delta t'$ by the more specific symbols T and T_0, respectively,

$$T = \gamma T_0 = \frac{T_0}{(1 - v^2/c^2)^{1/2}}. \tag{11.1}$$

We can deduce from this quite generally that *a clock moving uniformly with velocity v through an inertial frame S goes slow by a factor $(1 - v^2/c^2)^{1/2}$ relative to the standard clocks at rest in S.* Clearly, then, the fastest rate is ascribed to a clock in its rest frame, and this is called its *proper rate.* On the other hand, at speeds close to the speed of light, the rate of the clock would be close to zero.

Like length contraction, time dilation can be regarded as a 'velocity perspective' effect. Nothing at all happens to the moving clock itself: we need only move the eye along with it, and its rate returns to normal.

If an *ideal* clock moves *non-uniformly* through an inertial frame, we shall *assume* that acceleration as such has no effect on the rate of the clock, i.e. that its instantaneous rate depends only on its instantaneous speed v according to the above rule. This we call the *clock hypothesis.* It can also be regarded as the definition of an 'ideal' clock. By no means all clocks meet this criterion. For example, a spring-driven pendulum clock whose bob is connected by two coiled springs to the sides of the case (so that it works without gravity) will clearly increase its proper rate as it is accelerated 'upwards'; on the other hand, it is insensitive to sideways acceleration. But in any case, as stressed by Sexl, the absoluteness of acceleration ensures that ideal clocks *can* be built, in principle. We need only take an arbitrary clock, observe whatever effect acceleration has on it, then attach to it an accelerometer and a computer that continually makes allowance for the acceleration and 'corrects' the reading. By contrast, the velocity effect (11.1) *cannot* be eliminated. As we shall see, certain natural clocks (vibrating atoms, decaying muons) conform very accurately to the clock hypothesis. Generally this could be expected to happen if the clock's internal driving forces greatly exceed the accelerating force.

For the non-uniform motion of an *infinitesimal* rigid body a similar assumption is usually made, namely that the relation between its length in the direction of motion and its proper length depends only on its instantaneous velocity v according to equation (9.1). This we call the *length hypothesis*.

Time dilation, like length contraction, must *a priori* be symmetric: if one inertial observer considers the clocks of a second inertial observer to run slow, the second must also consider the clocks of the first to run slow. Figure 8 shows in detail how this happens. Synchronized standard clocks A, B, C, . . . and A′, B′, C′, . . . are fixed at certain equal intervals along the x-axes of two frames S and S′ in standard configuration. The figure shows three world-maps made, at convenient equal time intervals, in a third frame S″ relative to which S and S′ have equal and opposite velocities. In each world-map the clocks of S and S′ will all be seen to indicate different times, since simultaneity is relative. Suppose the clocks in the diagram indicate seconds. As can be seen, A′ reads 4 seconds ahead of A in Fig. 8(a), only 2 seconds ahead of C in Fig. 8(b), and equal with E in Fig. 8(c). Thus A′ loses steadily relative to the clocks in S. Similarly E loses steadily relative to the clocks in S′, and indeed all clocks in the diagram lose at the same rate relative to the clocks of the other frame.

Figure 8 refers to essentially the same situation as Fig. 6. (Note how the left end-points of the 'rods' AE and A′E′ meet before the right end-

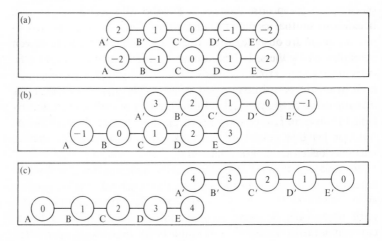

F<small>IG</small>. 8

points in S, and vice versa in S'.) The figure well illustrates how time dilation is related to the relativity of simultaneity.

Time dilation, like length contraction, is *real*. And this *has* been confirmed experimentally. For example, certain mesons (muons) reaching us from the top of the atmosphere (where they are produced by incoming cosmic rays), are so short-lived that, even had they travelled at the speed of light, their travel time in the absence of time dilation would exceed their lifetime by factors of the order of 10. Rossi and Hall in 1941 timed such muons between the summit and foot of Mt Washington, and found their lifetimes were indeed dilated in accordance with (11.1). Equivalent experiments with muons circling 'storage rings' (with velocities corresponding to $\gamma \approx 29$) at the CERN laboratory in 1975[1] have refined these results to an impressive accuracy of ~ 1 per cent, and have additionally shown that, to such accuracy, accelerations of up to $10^{18} g$ (!) do not contribute to the muons' time dilation. It may be objected that muons are not clocks—but the time dilation argument applies to *any* temporal change or process and therefore also to muon decay and even human lifetimes (as, for example, that of astronauts on fast space journeys). To see this, we need only imagine a standard clock to travel *with* the muon or the space traveller.

Another striking instance of time dilation is provided by 'relativistic focusing' of electrically charged particles, which plays a role in the operation of high-energy particle accelerators. Any stationary cluster of electrons (or protons) tends to expand at a characteristic rate because of mutual electrostatic repulsion. But electrons in a fast-moving beam are observed to spread at a much slower rate. (This can be explained by the same mechanism as that which causes parallel currents to attract each other.) If we regard the stationary cluster as a kind of clock, we have here an almost visible manifestation of the slowing down of a moving clock. Yet another manifestation, the so-called transverse Doppler effect, will be discussed in Section 17. (It, too, has led to a confirmation of the clock hypothesis for certain natural clocks.)

Nowadays, amazingly, time dilation can even be observed in macroscopic clocks. This was first done in 1971—though only to an accuracy of about 10 per cent—by Hafele and Keating,[2] who simply took some very accurate caesium clocks around the world on commercial airliners! Later (in 1975 and 1976) these experiments were refined by C. O. Alley and coworkers from the University of

Maryland, who placed caesium clocks in specially equipped military airplanes flying 30-mile circuits at about 300 mph above the Chesapeake Bay for 15 hours at a time. Among other sophisticated techniques, laser pulses were used for remote time control, and an overall accuracy of about 1 per cent was achieved.[3]

[1] Bailey, J. *et al.* (1975) *Physics Letters* **55B**, 420. See also Bailey, J. and Picasso, E. (1970) *Progress in Nuclear Physics*, **12**, 43.
[2] Hafele, J. C. and Keating, R. (1972) *Science* **177**, 166.
[3] Alley, C. O. (1979) in 33rd *Annual Frequency Control Symposium*, published by Electronic Industries Association, Washington, D. C.

12. The twin paradox

Like length contraction, so also time dilation can lead to an apparent paradox when viewed by two different observers. In fact this paradox, the so-called twin or clock paradox, or paradox of Langevin (1911), is the oldest of the relativistic paradoxes. It is quite easily resolved, but seems to possess some hidden emotional content that makes it the subject of interminable debate among dilettantes in relativity.

Consider two synchronized standard clocks A and B at rest at a point P of an inertial frame S. Let A remain at P while B is briefly accelerated to some constant velocity v with which it travels to a distant point Q in S. There it is decelerated briefly and made to return with velocity $-v$ to P. If, of two twins, one travels with B while the other remains with A, the B-twin will be younger than the A-twin when they meet again, for each ages at the same rate relative to *his* clock.

Now the paradox is this: cannot B (we shall identify clocks with persons) claim with equal right that it was *he* who remained where he was, while A went on a round-trip, and that consequently A should be the younger when they meet again? The answer is no, and this resolves the paradox: A has remained at rest in a single inertial frame while B was accelerated out of his rest frame at P, at Q, and once again at P. These accelerations are recorded on B's accelerometer and he can therefore be under no illusion that it was he who remained at rest, or that he and his twin entered this 'experiment' symmetrically. Of course, the two accelerations at P are not essential for the gist of the argument—age comparisons could be made in passing—but the acceleration at Q is vital.

Still, it could be argued that there *is* symmetry between A and B for 'most of the time', namely during the times of B's free fall. The three asymmetric accelerations can be confined to arbitrarily short periods (as measured by A—they are even shorter as measured by B). How is it then that a large asymmetric effect can build up, and, moreover, one that is proportional to the symmetric parts of the motion? But (as pointed out by Bondi) the situation is no more strange than that of two drivers A and B going from O to P to Q (three points in a straight line), A going directly, while B deviates at P to a point R off the line, and thence to Q. They behave quite similarly except that B turns his steering wheel and readjusts his speed briefly at P and again at R. Yet when they meet at Q, their odometers may indicate a large mileage difference!

In a way, the twins' eventual age difference can be seen to arise during B's initial acceleration away from P. During this period, however brief, if his γ-factor gets to be 2, say, B finds that he has accomplished more than half his outward journey! For he has transferred himself to a frame in which the distance between P and Q is halved (length contraction), and this halving is real to him in every way. Thus he accomplishes his outward trip in about half the time that A ascribes to it, and the same is true of his return trip.

Many arguments of this nature exist, which illuminate the lack of symmetry between the twins and demonstrate the self-consistency of the theory. [See, for example, Exercise III(1).] But the paradox is disposed of as soon as the asymmetry has been established. Sciama has made perhaps the most significant remark about this paradox: it has, he said, the same status as Newton's experiment with the two buckets of water—one, rotating, suspended below the other, at rest. If these were the whole content of the universe, it would indeed be paradoxical that the water surface in the one should be curved and that in the other flat. But inertial frames have a real existence too, and relative to the inertial frames there is no symmetry between the buckets, and no symmetry between the twins, either.

It should be noted, finally, that the clock paradox is entirely independent of the clock hypothesis. Whatever the effects of the accelerations as such may be on the moving clock or organism, these effects can be dwarfed by simply extending the periods of free fall.

13. Velocity transformation

Once again, let us consider two inertial frames S and S' in standard configuration. Let **u** be the vector velocity in S of a particle or simply of a geometrical point (so as not to exclude the possibility $u \geq c$). We wish to find the velocity $\mathbf{u}'$ of this point in S'. As in classical kinematics, we define

$$\mathbf{u} = (u_1, u_2, u_3) = (\mathrm{d}x/\mathrm{d}t, \mathrm{d}y/\mathrm{d}t, \mathrm{d}z/t) \tag{13.1}$$

$$\mathbf{u}' = (u'_1, u'_2, u'_3) = (\mathrm{d}x'/\mathrm{d}t', \mathrm{d}y'/\mathrm{d}t', \mathrm{d}z'/\mathrm{d}t'), \tag{13.2}$$

where the differentials refer to two successive events at the moving point. Substitution from (7.3) into (13.2), division of each numerator and denominator by $\mathrm{d}t$, and comparison with (13.1), now immediately yields the velocity transformation formulae:

$$u'_1 = \frac{u_1 - v}{1 - u_1 v/c^2}, \quad u'_2 = \frac{u_2}{\gamma(1 - u_1 v/c^2)}, \quad u'_3 = \frac{u_3}{\gamma(1 - u_1 v/c^2)}. \tag{13.3}$$

Since no assumption as to the uniformity of $\mathbf{u}$ was made, these formulae apply equally to the instantaneous velocity in a non-uniform motion. Note also how they reduce to the classical formulae (6.2) when either $v \ll c$, or $c \to \infty$ formally.

We can pass to the inverse relations without further effort by the standard method of interchanging primed and unprimed symbols and replacing v by $-v$. (For if we replace unprimed by primed, and primed by doubly primed symbols, we evidently get a transformation from S' to a frame S'' which moves with velocity v relative to S'; replacing v by $-v$ then makes S'' into S.) Thus,

$$u_1 = \frac{u'_1 + v}{1 + u'_1 v/c^2}, \quad u_2 = \frac{u'_2}{\gamma(1 + u'_1 v/c^2)}, \quad u_3 = \frac{u'_3}{\gamma(1 + u'_1 v/c^2)}. \tag{13.4}$$

These equations can be regarded as giving the resultant, $\mathbf{u}$, of the two velocities $\mathbf{v} = (v, 0, 0)$ and $\mathbf{u}'$, and are therefore occasionally referred to as the relativistic *velocity addition* formulae. In particular, the first member gives the resultant of two collinear velocities v and u'_1 and is therefore of the same form as the velocity parameter in the resultant of two successive Lorentz transformations [see property (vii) of Section 7].

Writing $u = (u_1^2 + u_2^2 + u_3^2)^{1/2}$ and $u' = (u_1'^2 + u_2'^2 + u_3'^2)^{1/2}$ for the magnitudes of corresponding velocities in S and S', we have, from (6.6) and (7.3)(i),

$$\mathrm{d}t^2 (c^2 - u^2) = \mathrm{d}t'^2 (c^2 - u'^2) = \mathrm{d}t^2 \gamma^2(v) (1 - u_1 v/c^2)^2 (c^2 - u'^2).$$

Rearranging this yields the following important relations, the last of which clearly remains valid even when the axes of S and S' are equally rotated:

$$c^2 - u'^2 = \frac{c^2(c^2 - u^2)(c^2 - v^2)}{(c^2 - u_1 v)^2} = \frac{c^2(c^2 - u^2)(c^2 - v^2)}{(c^2 - \mathbf{u} \cdot \mathbf{v})^2}. \quad (13.5)$$

It implies that if $u' < c$ and $v < c$, then $u < c$. Hence the resultant of two velocities less than c is always a velocity less than c. This shows that, however many velocity increments (less than c) a particle receives in its successive instantaneous rest frames, it can never attain the velocity of light. Thus the velocity of light plays the role in relativity of an infinite velocity, inasmuch as no 'sum' of lesser velocities can ever equal it. More generally, (13.5) shows that if $v < c$ (as is the case for any two inertial frames), $u \lesseqgtr c$ implies $u' \lesseqgtr c$, respectively, and vice versa.

Rewriting (13.5) in terms of $\gamma(u)$, $\gamma(u')$, $\gamma(v)$, we get an equation which, on taking square roots, yields the first of the following two useful relations,

$$\frac{\gamma(u')}{\gamma(u)} = \gamma(v)\left(1 - \frac{u_1 v}{c^2}\right), \quad \frac{\gamma(u)}{\gamma(u')} = \gamma(v)\left(1 + \frac{u'_1 v}{c^2}\right), \quad (13.6)$$

while the second results from the generally valid process of interchanging primed and unprimed symbols and replacing v by $-v$. [Cf. the remark before (13.4).] These relations show how the γ-factor of a moving particle transforms.

14. Transformation of linear acceleration

When a particle moves non-uniformly, it is useful to know how to transform not only its velocity, but also its acceleration. In this section we restrict our discussion to one-dimensional motion; the general case will be discussed, with more powerful mathematical tools, in Section 23.

Suppose, then, a particle P moves along the x-axis of an inertial frame S with varying velocity u. Let S' (in standard configuration with S) be the instantaneous rest frame of P at some instant $t = t_0$. Then $u = v$ and $u' = 0$ at t_0, but u and u' vary while v is constant. First note, from (7.3)(i), that

$$dt'/dt = \gamma(1 - uv/c^2).$$

Now differentiate (13.4)(i), with $u_1 = u$ and $u'_1 = u'$, at time t_0, i.e. put $v = u$ and $u' = 0$ *after* differentiation:

$$\frac{du}{dt} = \frac{du'}{dt} - \frac{u^2}{c^2}\frac{du'}{dt} = \frac{du'}{dt'}\left(1 - \frac{u^2}{c^2}\right)^{3/2}.$$

If we define the *proper acceleration* α of P as that which is measured in P's instantaneous rest frame, we can write our result in the form

$$\alpha = \left(1 - \frac{u^2}{c^2}\right)^{-3/2}\frac{du}{dt} = \gamma^3(u)\frac{du}{dt} = \frac{d}{dt}[\gamma(u)u]. \qquad (14.1)$$

Now forget that S′ denoted P's rest frame, and let S′ be *any* frame in standard configuration with S. For *it* we can prove a result analogous to (14.1), and so we have

$$\left(1 - \frac{u'^2}{c^2}\right)^{-3/2}\frac{du'}{dt'} = \left(1 - \frac{u^2}{c^2}\right)^{-3/2}\frac{du}{dt}. \qquad (14.2)$$

If need be, we could now express du'/dt' entirely in terms of S-quantities by use of (13.6)(i).

Under the Galilean transformation, the general acceleration is invariant: $d\mathbf{u}/dt = d\mathbf{u}'/dt'$. We note that in relativity this is no longer true.

A case of particular interest is that of rectilinear motion with *constant* proper acceleration α. We can then integrate (14.1) at once, choosing $t = 0$ when $u = 0$: $\alpha t = \gamma u$. Solving this equation for u, integrating once more, and suitably adjusting the constant of integration, yields the following equation of motion:

$$x^2 - c^2 t^2 = c^4/\alpha^2. \qquad (14.3)$$

Thus, for obvious reasons, rectilinear motion with constant proper acceleration is called *hyperbolic motion* (see Fig. 9). (The corresponding classical calculation gives $x = \frac{1}{2}\alpha t^2$, i.e. 'parabolic motion'.) Note that $\alpha = \infty$ implies $x = \pm ct$, hence the proper acceleration of a photon can be taken to be infinite. Note also, by reference to Fig. 9, that a photon emitted a distance c^2/α behind the particle when the latter is momentarily at rest, cannot catch up with it. (Its graph is the asymptote to the particle's hyperbola.)

Consider next the equation

$$x^2 - c^2 t^2 = X^2, \qquad (14.4)$$

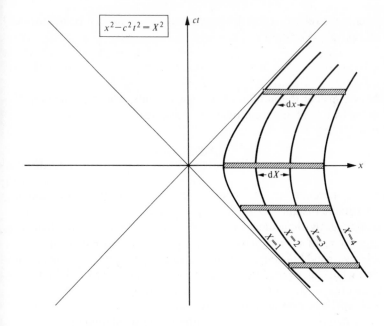

$$x^2 - c^2 t^2 = X^2$$

FIG. 9

for various values of the parameter X. For each fixed X it represents a particle moving with constant proper acceleration c^2/X. Altogether it represents, as we shall show, a rod 'moving rigidly' in the x-direction.

By the *rigid motion* of a body one understands a motion during which every small volume element of the body shrinks always in the direction of its motion in proportion to its instantaneous Lorentz factor relative to a given inertial frame. Thus every small volume element preserves its dimensions in its instantaneous rest frame, which shows that the definition is intrinsic, i.e. frame-independent. It also shows that during rigid motion no elastic stresses arise. A body moving rigidly cannot start to rotate, since circumferences of circles described by points of the body would have to shrink, while their radii would have to remain constant, which is impossible. In general, therefore, the motion of one point of a rigidly moving body determines that of all others. Note that the accelerating rod discussed at the end of (viii) of Section 7 moves rigidly in the present technical sense, even though it is bent in S.

Going back to equation (14.4), we can find (by implicit differentiation) the velocity u and the corresponding γ-factor of a point moving so that X = constant:

$$u = \frac{dx}{dt} = \frac{c^2 t}{x}, \quad \gamma(u) = \frac{x}{X}. \tag{14.5}$$

Now consider the motion of two such points, whose parameters X differ by dX; at any fixed time t we have, again from (14.4) and then from (14.5),

$$dx = \frac{X dX}{x} = \frac{dX}{\gamma(u)}. \tag{14.6}$$

Hence at every instant t = constant the two points are separated by a coordinate distance dx inversely proportional to their γ-factor, and consequently the element bounded by these points 'moves rigidly'; moreover, dX is recognized as its proper length. Since this applies to any two neighbouring points in the aggregate represented by (14.4), that whole aggregate 'moves rigidly', like an unstressed rod. Figure 9 shows the position of such a rod at various instants t = constant. The rod cannot be extended to negative values of X, since the asymptotes in the diagram represent photon paths; if continued right up to $X = 0$, the rod 'ends in a photon'.

Exercises II

1. An 18-foot pole, while remaining parallel to the x-axis, moves with velocity $(v, -w, 0)$ relative to frame S, where $\gamma(v) = 3$, and w is positive. The centre of the pole passes the centre of a 9-foot hole in a plate that coincides with the plane $y = 0$. Explain, from the point of view of the usual second frame S′ moving with velocity v relative to S, how the pole gets through the hole.

2. In the situation illustrated in Fig. 8, find the relative velocity between the frames S and S′, the distance between neighbouring clocks in either frame, and the relative velocity between S′ and S″. [*Answers*: $(2\sqrt{2}/3)c, \sqrt{2} \times 300\,000$ km, $\sqrt{2}c/2$.]

3. Two particles move along the x-axis of S at velocities $0.8\,c$ and $0.9\,c$, respectively, the faster one momentarily 1 m behind the slower one. How many seconds elapse before collision?

4. A rod of proper length 10 cm moves longitudinally along the x-axis of S at speed $\frac{1}{2}c$. How long (in S) does it take a particle, moving oppositely at the same speed, to pass the rod?

5. In a given inertial frame, two particles are shot out simultaneously from a given point, with equal speeds v, in orthogonal directions. What is the speed of each particle relative to the other? [*Answer*: $v(2 - (v^2/c^2))^{1/2}$.]

6. A rod, having slope m relative to the x-axis of S, moves along that axis at speed u. What is the rod's slope in S'? [*Answer*: $m\gamma(v)$ $(1 - uv/c^2)$.] *Note*: The reader should think of two ways to do this problem, one using length contraction, the other not.

7. Show that the result of relativistically 'adding' a velocity $\mathbf{u}'$ to a velocity $\mathbf{v}$ is not, in general, the same as that of 'adding' a velocity $\mathbf{v}$ to a velocity $\mathbf{u}'$. [*Hint*: consider $\mathbf{u}' = (0, u', 0)$ and $\mathbf{v} = (v, 0, 0)$.] Also show that the *magnitudes* of these two 'sums' are always the same. [*Hint*: if the velocity $\mathbf{v}$ of S' relative to S is not along the x-axis, consider first a standard Lorentz transformation and then rotate the axes of S and S' *equally* to give $\mathbf{v}$ its desired direction; only then does it make sense to 'add' $\mathbf{u}'$ to $\mathbf{v}$. Cf. (13.5).]

8. The *rapidity* ϕ, of a particle moving with velocity u, is defined by $\phi = \tanh^{-1}(u/c)$ [cf. Exercise I (12)]. Prove that *collinear* rapidities are additive, i.e. if A has rapidity ϕ relative to B, and B has rapidly ψ relative to C, then A has rapidity $\phi + \psi$ relative to C.

9. How many successive velocity increments of $\frac{1}{2}c$ from the instantaneous rest frame are needed to produce a resultant velocity of (i) $0.99\,c$, (ii) $0.999\,c$? [*Answer*: 5, 7. *Hint*: $\tanh 0.55 = 0.5$, $\tanh 2.65 = 0.99$, $\tanh 3.8 = 0.999$.]

10. If $\phi = \tanh^{-1}(u/c)$, and $e^{2\phi} = z$, prove that n consecutive velocity increments u from the instantaneous rest frame produce a velocity $c(z^n - 1)/(z^n + 1)$.

11. In S' a particle is momentarily at rest and has acceleration α in the y'-direction. What is the magnitude and direction of its acceleration in S? [*Hint*: use an argument analogous to that leading to (13.6).] In a muon 'storage ring' of radius 7 m at the CERN laboratories in 1975, muons circled around at a speed $v = 0.9994\,c$. Find the magnitude of their proper acceleration.

12. A certain piece of elastic breaks when it is stretched to twice its unstretched length. At time $t = 0$, all points of it are accelerated longitudinally with constant proper acceleration α, from rest in the unstretched state. Prove that the elastic breaks at $t = \sqrt{3}c/\alpha$.

13. In a frame S consider a rectilinearly moving particle having velocity u, rapidity ϕ, and proper acceleration α, and let τ be the proper time elapsed at a clock carried by the particle. Prove that

$d\phi/d\tau = \alpha/c$. [*Hint*: Exercise 8 above.]

14. In the situation of the preceding exercise, consider the special case of motion along the x-axis with $\alpha = $ constant (hyperbolic motion). If $\tau = 0$ and $u = 0$ when $t = 0$, establish the following formulae:

$$u/c = \tanh(\alpha\tau/c), \qquad \gamma(u) = \cosh(\alpha\tau/c),$$

$$\alpha t/c = \sinh(\alpha\tau/c), \qquad \alpha x/c^2 = \cosh(\alpha\tau/c).$$

15. Given that g, the acceleration of gravity at the earth's surface, is $\sim 9.8\,\mathrm{m\,s^{-2}}$, and that a year has $\sim 3.2 \times 10^7$ seconds, verify that, in units of years and light years, $g \approx 1$. A rocket moves from rest in an inertial frame S with constant proper acceleration g (thus giving maximum comfort to its passengers). Find its Lorentz factor relative to S when its own clock indicates times $\tau = 1$ day, 1 year, 10 years. Find also the corresponding distances and times travelled in S. If the rocket accelerates for 10 years of its own time, then decelerates for 10 years, and then repeats the whole manoeuvre in the reverse direction, what is the total time elapsed in S during the rocket's absence? [*Answers*: $\gamma = 1.000\,0038$, 1.5431, 11013; $x = 0.000\,0038$, 0.5431, 11012 light years; $t = 0.0027$, 1.1752, 11013 years; $t = 44\,052$ years. To obtain some of these answers you will have to consult tables of $\sinh x$ and $\cosh x$. At small values of their arguments a Taylor expansion suffices.]

16. Evidently the equation of motion (14.4) of the uniformly accelerating rod (or 'rocket') is invariant under a standard Lorentz transformation. What is the significance of this invariance? Prove that for any event $\mathscr{P}$ on the rocket there exists a frame S' in which the spacetime diagram of the rocket looks exactly like Fig. 9 but with $\mathscr{P}$ on the x'-axis. Now prove that the 'radar distance' (the proper time of a 'light-echo' multiplied by $c/2$) of a point on the rocket at parameter X_2, from an observer riding on the rocket at parameter X_1, is $X_1 \sinh^{-1}[(X_2^2 - X_1^2)/2X_1X_2]$. [*Hint*: choose the reflection event on the x-axis.]

17. In a given frame, a particle A moves hyperbolically with proper acceleration α from rest at $t = 0$. At $t = 0$ a photon B is emitted in the same direction, a distance c^2/α behind A. Prove that in A's instantaneous rest frames the distance AB is always c^2/α.

18. Two inertial frames S and S' are in standard configuration while a third, S'', moves with velocity v' along the y'-axis, its axes parallel to

those of S'. If the line of relative motion of S and S'' makes angles θ and θ'' with the x and x''-axes, respectively, prove that $\tan \theta = v'/v\gamma(v)$, $\tan \theta'' = v'\gamma(v')/v$. [*Hint*: use (13.4).] The inclination $\delta\theta$ of S'' relative to S is defined as $\theta'' - \theta$. If $v, v' \ll c$, prove that $\delta\theta \approx vv'/2c^2$. If a particle describes a circular path at uniform speed $v \ll c$ in a given frame S, and consecutive instantaneous rest frames, say S' and S'', always have zero relative inclination, prove that after a complete revolution the instantaneous rest frame is tilted through an angle $\pi v^2/c^2$ in the sense opposite to that of the motion. ['*Thomas precession*.' *Hint*: Let a tangent and radius of the circle coincide with the x'- and y'-axes. Then $v' = v^2 \, dt/a, a = $ radius.]

III
RELATIVISTIC OPTICS

15. Introduction

In this chapter we are concerned with some aspects of the propagation of light. We shall here take a very elementary and pragmatic approach, sometimes treating light as waves, sometimes as particles (photons). In order to discuss frequency, we shall regard atoms as emitting light in a regular series of 'pulses' (corresponding to wavecrests). Later we shall find that these simple ideas are compatible with the full electromagnetic theory of light.

16. The drag effect

Relativity provided an ideally simple solution to a problem that had considerably exercised the ingenuity of theoreticians before. The question is to what extent a flowing liquid will 'drag' light along with it. Flowing air, of course, drags sound along totally, but the optical situation is different: on the basis of an ether theory, it would be conceivable that there is no drag at all, since light is a disturbance of the ether and not of the liquid. Yet experiments indicated that there *was* a drag: the liquid seemed to force the ether along with it, but only partially. If the speed of light in the liquid *at rest* is u', and the liquid is set to move with velocity v, then the speed of light relative to the outside was found to be of the form

$$u = u' + kv, \quad k = 1 - 1/n^2, \tag{16.1}$$

where k is the 'drag coefficient', a number between zero and one indicating what fraction of its own velocity the liquid imparts to the ether within, and n is the refractive index c/u' of the liquid. Fifty years before Einstein, Fresnel succeeded in giving a plausible ether-based explanation of this. From the point of view of special relativity, however, the result (16.1) is nothing but the relativistic velocity addition formula! The light travels relative to the liquid with velocity u', and the liquid travels relative to the observer with velocity v, and therefore [cf. (13.4)(i)]

$$u = \frac{u' + v}{1 + u'v/c^2} = \frac{(c/n) + v}{1 + v/cn} \approx c\left(\frac{1}{n} + \frac{v}{c}\right)\left(1 - \frac{v}{cn}\right)$$

$$\approx u' + v\left(1 - \frac{1}{n^2}\right), \qquad (16.2)$$

neglecting terms of order v^2/c^2 in the last two steps. Einstein already gave the velocity addition formula in his 1905 paper, but it took two more years before Laue made this beautiful application of it.

17. The Doppler effect

Waves from an approaching light-source have higher frequency than waves from a stationary source. In the frame of the source, this is because the observer moves *into* the wave train, and in the frame of the observer it is because the source, chasing its waves, bunches them up. The opposite happens when the source recedes. Similar effects exist for sound and other wave phenomena; all are named after the Austrian physicist Doppler.

Classically, it made a difference whether the observer or the light-source, or neither, were at rest in the ether. In relativity, of course, all that matters is their relative motion. But relativity has also added another modification: the time dilation of the source (or of the observer).

Let a light-source P travelling through an inertial frame S have instantaneous velocity **u**, and radial velocity component u_r relative to the origin-observer O [see Fig. 10(a)]. Let the time between successive

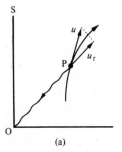

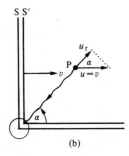

(a)   (b)

Fig. 10

pulses be dt_0 as measured by a comoving observer *at* the source, and therefore $dt_0\gamma(u)$ in S (by time dilation). In that time, the source has increased its distance from O by $dt_0\gamma(u)u_r$. Consequently these pulses arrive at O a time $dt = dt_0\gamma + dt_0\gamma u_r/c$ apart. But dt_0 and dt are inversely proportional, respectively, to the proper frequency v_0 of the source and its frequency v as observed by O. So

$$\frac{v_0}{v} = \frac{1 + u_r/c}{(1 - u^2/c^2)^{1/2}} = 1 + \frac{u_r}{c} + \frac{1}{2}\frac{u^2}{c^2} + O\left(\frac{u^3}{c^3}\right). \tag{17.1}$$

Our series expansion separates the 'pure' Doppler effect $1 + u_r/c$ from the contribution $\frac{1}{2}u^2/c^2$ of time dilation, to the order shown. The prerelativistic formula had the Lorentz factor missing, but it was considered valid only for an observer at rest in the ether.

Note that a real light-source is usually made up of many atoms or molecules and may emit light at many frequencies. But all obey the same law, and thus the entire spectrum of the source is 'Doppler-shifted'. Note also that the above argument and formula apply equally to the visually observed frequency v of an arbitrarily moving ideal *clock* of proper frequency v_0.

When the motion of the source is purely radial, $u_r = u$ and equation (17.1) reduces to

$$\frac{v_0}{v} = \left(\frac{1 + u/c}{1 - u/c}\right)^{1/2}. \tag{17.2}$$

It is also useful to have a formula relating the frequencies v and v' ascribed by *two* observers O and O' at the *same* event to an incoming ray of unspecified origin. Let O and O' be associated with the usual frames S and S', respectively. Let α be the angle which the *negative* direction of the ray makes with the x-axis of S. The trick here is to assume, obviously without loss of generality, that the ray originated from a source at rest in S'. Then (17.1) applies with the following specializations: $v_0 = v'$, $u = v$, $u_r = v\cos\alpha$ [see Fig. 10(b)]. So

$$\frac{v'}{v} = \frac{1 + (v/c)\cos\alpha}{(1 - v^2/c^2)^{1/2}}. \tag{17.3}$$

This formula allows us, when convenient, to evaluate the Doppler ratio in one inertial frame and then transform it to the frame of interest. This is just what we would do if, for example, the source were at rest in an inertial frame through which the *observer* moves non-uniformly. In such cases, however, we need another conceptual tool:

the principle that *an accelerating observer makes the same local time and distance measurements as an inertial observer momentarily comoving with him.* This follows from the length and clock hypotheses made in Section 11.

A simple case in point is the frequency shift between a source at the centre of a rapidly turning rotor, and an 'observer' (a piece of apparatus) at the rim, which moves, let us say, with linear velocity v. Setting $\alpha = 90°$ and $v = v_0$ in (17.3), we find $v' = v_0 \gamma(v)$ for the observed frequency of a source of proper frequency v_0. This, of course, is entirely due to the time dilation of the moving observer. The experiment was performed—with a view to demonstrating such time dilation—by Hay, Schiffer, Cranshaw, and Engelstaff in 1960, using Mössbauer resonance. Agreement with the theoretical predictions was obtained to within an expected experimental error of a few per cent. Moreover, the experiment furnished some validation of the clock hypothesis, for the 'clock' at the receiver was clearly accelerated (up to about $6 \times 10^4 \, g$, in fact), and no measurable effect of this acceleration could be detected.

Time dilation is the only cause of the frequency shift whenever there is no radial motion of the source or the observer. This is the so-called *transverse Doppler effect*, and has long been considered as a possible basis for time dilation experiments. Prior to the rotor experiments, however, it was difficult to ensure exact transverseness in the motion of the sources (e.g. fast-moving hydrogen ions). The slightest radial component would swamp the transverse effect. Ives and Stilwell in 1938 cleverly used a to-and-fro motion of ions whereby the first-order Doppler effect cancelled out, and only the contribution of time dilation remained, which they were able to measure to an accuracy of about 10 per cent. Theirs was a historic experiment, being the first to demonstrate the reality of time dilation. (Yet by a curious irony of fate, Ives and Stilwell were lone hold-outs for Lorentz's ether theory and rejected special relativity.)

A similar cancelling of the first-order contribution occurs in the so-called *thermal Doppler effect*. Radioactive nuclei bound in a hot crystal move thermally in a rapid and random way. Because of this randomness, their first-order (classical) Doppler effects average out, but not the second-order (relativistic) time dilation effects. The former cause a mere broadening of the spectral lines, the latter a shift of the entire spectrum. This shift was observed, once again by use of Mössbauer resonance, in 1960 by Rebka and Pound. The accuracy of that experiment was also of the order of 10 per cent. However, a by-product of the experiment was an impressive validation of the clock hypothesis: in spite of proper accelerations up to $10^{16}\,g(!)$, these nuclear 'clocks' were slowed simply by the velocity factor $(1 - v^2/c^2)^{1/2}$.

18. Aberration and the visual appearance of moving objects

Anyone who has driven in the rain or snow knows that it seems to come at one obliquely. For similar reasons, if two observers measure the angle which an incoming ray of light makes with their relative line of motion, their measurements will generally not agree. This phenomenon is called aberration, and of course it was well known before relativity. Nevertheless, as in the case of the Doppler effect, the relativistic formula contains a correction, and it applies to all pairs of observers, whereas the prerelativistic formula was simple only if one of the observers was at rest in the ether frame.

To obtain the basic aberration formulae, consider an incoming light signal whose negative direction makes angles α and α' with the x-axes of the usual two frames S and S', respectively [cf. Fig. 10(b)]. The velocity transformation formula (13.3)(i) can evidently be applied to this signal, with $u_1 = -c\cos\alpha$ and $u_1' = -c\cos\alpha'$, yielding

$$\cos\alpha' = \frac{\cos\alpha + v/c}{1 + (v/c)\cos\alpha}.\tag{18.1}$$

Similarly, from (13.3)(ii) we obtain the alternative formula (assuming temporarily, without loss of generality, that the signal lies in the xy-plane)

$$\sin\alpha' = \frac{\sin\alpha}{\gamma[1 + (v/c)\cos\alpha]}.\tag{18.2}$$

But the most interesting version of the aberration formula is obtained by substituting from equations (18.1) and (18.2) into the trigonometric identity

$$\tan\tfrac{1}{2}\alpha' = \sin\alpha'/(1 + \cos\alpha'),$$

which gives

$$\tan\tfrac{1}{2}\alpha' = \left(\frac{c-v}{c+v}\right)^{1/2}\tan\tfrac{1}{2}\alpha.\tag{18.3}$$

For rays going *out* at angles α and α', we merely replace c by $-c$ in all the above formulae.

Aberration implies, for example, that as the earth travels along its orbit, the apparent directions of the fixed stars trace out small ellipses in the course of each year (with major axes of about 41 seconds of arc). Aberration also causes certain distortions in the visual appearance of extended uniformly moving objects. For, from the viewpoint of the rest frame of the object, as the observer moves past the conical pattern of rays converging from the object to the observer's eye, rays from its different points are unequally aberrated. Alternatively, from the viewpoint of the observer's rest frame, the light from different parts of the moving object takes different times to reach his eye, and thus it was emitted at different past times; the more distant points of the object consequently appear displaced relative to the nearer points in the direction opposite to the motion.

It is in connection with the visual appearance of uniformly moving objects that the relativistic results are a little unexpected. Following an ingenious argument of R. Penrose, let us draw a sphere of unit diameter around each observer's space-origin (see Fig. 11), cutting the negative and positive x-axis at points P and Q, respectively. All that an observer sees at any instant can be mapped onto this sphere (his 'sky'). Let it further be mapped from this sphere onto the tangent plane at Q (his 'screen') by stereographic projection from P. We recall that the angle subtended by an arc of a circle at the circumference is half of the angle subtended at the centre, and we have accordingly labelled the diagram (for a single incident ray). Thus the significance of (18.3) is seen to be precisely this: whatever the two momentarily coincident observers see, the views on their 'screens' are identical except for scale.

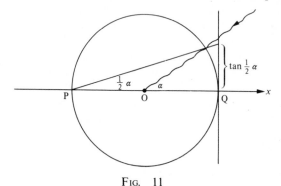

F<small>IG</small>.   11

Now consider, for example, a solid sphere Z at rest somewhere in the frame of an inertial observer O'. He sees a circular outline of Z in his sky, and projects a circular outline on his screen (for under stereographic projection, circles on the sphere correspond to circles or straight lines on the plane). Relative to the usual second observer O, of course, Z moves. Nevertheless, according to our theory, his screen image will differ from that of O' only in size, and thus it will also be circular; consequently, his 'sky' image of Z must be circular too. This shows that a moving sphere presents a circular outline to *all* observers *in spite* of length contraction! (Or rather: *because* of length contraction; for without length contraction the outline would be distorted.) By the same argument, moving straight lines (rods) will, in general, have the appearance of circular arcs, and flying saucers or fast-moving bicycle wheels are liable to look boomerang-shaped.

Another interesting though less realistic way of studying the visual appearance of moving objects is by use of what we may call 'supersnapshots'. These are life-size snapshots made by receiving *parallel* light from an object and catching it directly on a photographic plate held at right angles to the rays. One could, for example, make a supersnapshot of the *outline* of an object by arranging to have the sun behind it and letting it cast its shadow onto a plate. Moreover, what the eye sees (or the ordinary camera photographs) of a *small and distant* object approximates quite closely to a supersnapshot. Now, the surprising result (due to Terrell) is this: all supersnapshots that can be made of a uniformly moving object at a certain place and time by observers in any state of uniform motion are identical. In particular, they are all identical to the supersnapshot that can be made in the rest frame of the object.

To prove this result,[1] let us consider two photons P and Q travelling abreast along parallel straight paths a distance Δr apart, relative to some frame S. Let us consider two arbitrary events $\mathscr{P}$ and $\mathscr{Q}$ at P and Q, respectively. If $\mathscr{Q}$ occurs a time Δt after $\mathscr{P}$, then the space separation between $\mathscr{P}$ and $\mathscr{Q}$ is evidently $(\Delta r^2 + c^2 \Delta t^2)^{1/2}$, and thus, by (6.9), the squared interval between $\mathscr{P}$ and $\mathscr{Q}$ is $-\Delta r^2$, independently of their time separation. But if, instead of travelling abreast, Q leads P by a distance Δl, then the space separation between $\mathscr{P}$ and $\mathscr{Q}$ would be $[\Delta r^2 + (c\Delta t + \Delta l)^2]^{1/2}$, and the squared interval would *not* be independent of Δt. Now, since squared interval is an invariant (and since parallel rays transform into parallel rays), it follows that two photons travelling abreast along parallel paths a distance Δr apart in one frame do precisely the same in all other frames. But a supersnapshot results from catching an array of photons travelling abreast along parallel paths on a plate orthogonal to those paths. By our present result, these photons travel abreast along parallel lines with the same space separation in *all* inertial frames, and thus the equality of super-snapshots is established.

Suppose, for example, the origin-observer O in S sees at $t = 0$ a small object, apparently on his y-axis ($\alpha = 90°$). Suppose this object is at rest in the usual second frame S'. The origin-observer O' in S' will see the object at an angle $\alpha < 90°$, given by (18.3). If the object is a cube with its edges parallel to the coordinate axes of S and S', O' of course sees the cube not face-on but rotated. The surprising result is that O, who might have expected to see a contracted cube face-on, also sees an uncontracted rotated cube! (However, the basic rotation effect is not specifically relativistic: classically there would also be rotation, though with distortion.)

---

[1] The proof given here was suggested by one used by V. F. Weisskopf in *Physics Today*, Sept. 1960, p. 24.

Exercises III

1. If the twins A and B, in the twin-paradox 'experiment' discussed on p. 34, visually observe the regular ticking of each other's standard clocks, describe quantitatively what each sees as B travels to a distant point Q and back. [*Hint*: draw a spacetime diagram, treating B's velocity changes as instantaneous.] Note that B receives slow ticks for half the time and fast ticks for the other half, whereas A receives slow ticks for *more* than half the time: hence A receives fewer ticks, hence B is younger when they meet again. This is one of the arguments often used to illustrate the 'non-paradoxicality' of the paradox. Where does relativity come into it—i.e. why does it fail classically?

2. Fill in the details of the following argument (due to H. Bondi) which yields the formulae for time dilation, the one-dimensional Doppler effect, and the one-dimensional velocity transformation, directly from the axioms. [*Hint*: a spacetime diagram may be found helpful.] Two observers P, Q are at rest in an inertial frame while a third, R, travels at uniform velocity from P to Q. Light of frequency v is emitted by P, which R partly reflects to P and partly transmits to Q. If R sees P's light Doppler-shifted by a factor $D(= v/v_R,$

v_R = frequency seen by R), P will see the reflected light Doppler-shifted by a factor D^2 (why?). Suppose P so times his emission that R receives light only between P and Q, and let P and R ascribe times t_0 and t_1, respectively, to this journey. Then P and Q receive light during times $t_0(1 + v/c)$ and $t_0(1 - v/c)$, respectively, and R receives $t_1 v/D$ complete waves. Thus

$$t_0\left(1 - \frac{v}{c}\right) = \frac{t_1 v}{D}\frac{1}{v} = \frac{t_1}{D}, \quad t_0\left(1 + \frac{v}{c}\right) = \frac{t_1 v D^2}{D}\frac{1}{v} = t_1 D.$$

Multiplying and dividing these equations gives two of the required results. If T is another collinear inertial observer having Doppler shifts D' and D'' relative to R and P respectively, justify $DD' = D''$ and deduce the relativistic velocity transformation. Where exactly do Einstein's two axioms enter the above argument?

3. A large disc rotates at uniform angular speed ω in an inertial frame S. Two observers O_1 and O_2 ride on the disc at radial distances r_1 and r_2 from the centre (not necessarily on the same radius). They carry clocks C_1 and C_2 which they adjust so that they keep time with the clocks in S, i.e. they speed up their natural rates by the Lorentz factors $\gamma_1 = (1 - r_1^2\omega^2/c^2)^{-1/2}, \gamma_2 = (1 - r_2^2\omega^2/c^2)^{-1/2}$, respectively. By the stationary nature of the situation, C_2 cannot appear to gain or lose relative to C_1. Deduce that, when O_2 sends a light-signal to O_1, this signal suffers a Doppler shift $v_2/v_1 = \gamma_2/\gamma_1$. Note, in particular, that there is no relative Doppler shift between any two observers equidistant from the centre, and give an alternative direct proof of this.

4. A source of light moves with speed v along the y-axis, and an observer moves with speed v along the x-axis. The source, when it emits the signal, is as far from the origin as the observer is when he receives it. What Doppler shift does he see? [Answer: $(\sqrt{2c} + v)/(\sqrt{2c} - v)$.]

5. On the hyperbolically moving rod discussed in Section 14, a light-signal is sent from a source at rest on the rod at $X = X_1$ to an observer at rest on the rod at $X = X_2$. Prove that the Doppler shift v_1/v_2 in the light is given by X_2/X_1. [Hint: refer to Fig. 9, and transform to a frame in which the observer is at rest at reception.]

6. From (17.3) and (18.2) derive the following interesting relation between Doppler shift and aberration: $v'/v = \sin\alpha/\sin\alpha'$.

7. Let Δt and $\Delta t'$ be the time separations in the usual two frames S and S' between two events occurring at a freely moving photon. If the

photon has frequencies v and v' in these frames, prove that $v/v' = \Delta t/\Delta t'$. [*Hint*: use the result of the preceding exercise.]

8. By combining the Doppler formula (17.3) with its inverse [obtained by interchanging primed and unprimed symbols and writing $-v$ for v—see before (13.4)], reobtain the aberration formula (18.1).

9. A rocket ship flies at a velocity v through a large circular hoop of radius a, along its axis. How far *beyond* the hoop is the rocket ship when the hoop appears exactly lateral to the pilot? [*Note*: this 'hindsight' effect *is not* specifically relativistic, although the exact result *is*.]

10. A source of light is fixed in S′ and in that frame it emits light uniformly in all directions. Show that for large v, the light in S is concentrated in a narrow forward cone; in particular, half the photons are emitted into a cone whose semi-angle is given by $\cos \theta = v/c$. This is called the 'headlight effect'. [*Hint*: consider the totality of rays leaving the source in S′.] Is the situation essentially different in the classical theory?

11. Two momentarily coincident observers travel towards a small and distant object. To one observer that object looks twice as large (linearly) as to the other. Prove that their relative velocity is $3c/5$.

12. Show that the ratio of the solid angles subtended in S and S′ by a thin pencil of light-rays converging on the coincident origins of these frames, their negative directions making angles α, α' with the respective x-axes, is given by

$$\frac{d\Omega}{d\Omega'} = \left(\frac{d\alpha}{d\alpha'}\right)^2 = \gamma^2(v)\left(1 + \frac{v}{c}\cos \alpha\right)^2 = \frac{v'^2}{v^2}.$$

[*Hint*: without loss of generality, consider a solid angle with circular normal cross section—cf. the argument associated with Fig. 11.]

13. A particle moves uniformly in a frame S with velocity $\mathbf{u}$ making an angle α with the positive x-axis. If α' is the corresponding angle in the usual second frame S′, prove the '*particle aberration formula*'

$$\tan \alpha' = \frac{\sin \alpha}{\gamma(v)(\cos \alpha - v/u)},$$

and compare this with (18.1) and (18.2). [*Hint*: use the velocity transformation formula as for (18.1) and (18.2).]

14. In a frame S, consider the equation $x \cos \alpha + y \sin \alpha = ut$. For fixed α it represents a plane propagating in the direction of its normal

with speed u, that direction being parallel to the xy-plane and making an angle α with the positive x-axis. We can evidently regard this plane as a wave front. Now transform x, y, and t directly to the usual frame S'. From the resulting equation deduce the following aberration formula for the wave normal:

$$\tan \alpha' = \frac{\sin \alpha}{\gamma(v)(\cos \alpha - uv/c^2)}.$$

By comparison with the result of the preceding exercise, note that waves and particles travelling with the same velocity 'aberrate' differently, unless that velocity is c. But waves with velocity $w = c^2/u$ aberrate like particles with velocity u—a result that will be of interest to us later.

15. A plane mirror moves in the direction of its normal with uniform velocity v in a frame S. A ray of light of frequency v_1 strikes the mirror at an angle of incidence θ, and is reflected with frequency v_2 at an angle of reflection ϕ. Prove that $\tan \frac{1}{2}\theta / \tan \frac{1}{2}\phi = (c+v)/(c-v)$ and

$$\frac{v_2}{v_1} = \frac{\sin \theta}{\sin \phi} = \frac{c \cos \theta + v}{c \cos \phi - v} = \frac{c + v \cos \theta}{c - v \cos \phi}.$$

These results are of some importance in thermodynamics. [*Hint*: let the mirror be fixed in S' and write down the obvious relations in that frame; then transform to S.]

16. A cube with its edges parallel to the coordinate axes moves with Lorentz factor 3 along the x-axis of an inertial frame S. A 'supersnapshot' of this cube is made in a plane $z = $ constant by means of light-rays parallel to the z-axis. Make an exact scale drawing of this supersnapshot.

17. Uniform parallel light is observed in two arbitrary inertial frames, say the usual frames S and S', with the light *not* parallel to the x-axes. If v and v' are the respective frequencies of the light in S and S', prove that the ratio $\rho:\rho'$ of the respective photon densities (number of photons per unit volume) is $v:v'$. [*Hint*: supersnapshots.]

18. Make a rough sketch, using the construction of Fig. 11, to show how some inertial observers could see the outline of a fast-moving disc (or 'flying saucer') boomerang-shaped.

IV

SPACETIME

19. Introduction

We have now reached a point at which further progress can be greatly clarified and simplified by the introduction of a new concept of space and time and a corresponding mathematical technique ideally suited for the discussion of special relativity. The new concept is Minkowski's four-dimensional '*spacetime*', and the mathematical technique is a tensor calculus—the theory of '*four-tensors*'—adapted specifically to that spacetime. The basic tensor notation, definitions and results that we shall use here can be found in the Appendix at the end of the book, and should be learned or reviewed—as the case may be—before the remainder of this chapter is read. Numbered references to the Appendix are prefixed with the letter 'A'.

20. Spacetime and four-tensors

The basic reference frames of special relativity are the inertial frames, each of which can be pictured as a swarm of freely moving test-particle-clocks, maintaining their mutual distances for ever. Although all types of coordinates could be used within each inertial frame, we have seen how convenient it is to adhere to 'standard' coordinates, i.e. spatial coordinates which are right-handed orthogonal Cartesians all based on a universal standard of length, and a time coordinate which makes the physics isotropic and is based on a universal standard of time. Such coordinates in all inertial frames will continue to be presupposed. But often we shall not need to make any further supposition about the *relative* configuration of the spatial axes, or about the choice of time-origins, when dealing with sets of inertial frames. If these axes and origins are chosen arbitrarily, the relations between the coordinates in two frames will be more complicated than in (6.14), though they will still be linear, as we have seen at the end of Section 6. Their exact form, which fortunately is rarely required, could always be obtained from (6.14) by substituting for the 'standard-configuration' coordinates in terms of the arbitrarily oriented coordinates envisaged here, and then solving for one set of these. The resulting transformation, we recall, is termed a Poincaré transformation.

As we have seen in Section 6, Poincaré transformations leave invariant the *differential squared interval*

$$ds^2 = c^2 dt^2 - dx^2 - dy^2 - dz^2 \qquad (20.1)$$

between neighbouring events separated by coordinate differentials dt, dx, dy, dz. This is analogous to the invariance of the differential squared *distance* $dx^2 + dy^2 + dz^2$ under rotations and translations of the orthogonal Cartesian axes in Euclidean three-space E_3. It suggests that the space of 'events' (t, x, y, z) (i.e. the entire world in space and time) might bear certain similarities to the Euclidean space of points (x, y, z)—each inertial standard coordinate system playing a role analogous to a set of three orthogonal Cartesian axes in E_3. The full recognition and development of this analogy is due to Minkowski (1908). Struck by the grandeur of his discovery, he exclaimed: 'Henceforth space by itself, and time by itself, are doomed to fade away into mere shadows, and only a kind of union of the two will preserve an independent reality.' Thus Minkowski may well be regarded as the father of the 'fourth dimension'—though, as with so many other early ideas in relativity, the great Poincaré had already anticipated some of this, without, however, nurturing it to full fruition.

According to the relativity principle, the laws of physics must have the same form in all inertial frames. Our attention is therefore naturally drawn to tensor equations as embodiments of physical laws, since they have the very property of being either true or false independently of the coordinate system. (Consider such equations as 'force equals mass times acceleration', $\mathbf{f} = m\mathbf{a}$, which are independent of the coordinates in E_3.) Now tensors are objects defined on a space V_N, either under *all* non-singular transformations of its coordinates or only under a certain subgroup of 'permissible' transformations. The second of these alternatives is of interest to us now: objects defined on the four-space of events, coordinatized by

$$x^0 = ct, \quad x^1 = x, \quad x^2 = y, \quad x^3 = z, \qquad (20.2)$$

which behave as tensors (see Section A5) under the Poincaré transformation group, will be called *four-tensors*. (The prefix 'four-' will often be omitted when there is no danger of confusion.) Note that from now on we take ct rather than just t as our first variable, in accordance with usage and convenience; and we number our indices 0, 1, 2, 3 rather than 1, 2, 3, 4. We shall use Greek indices $\mu, \nu \ldots$ for the range 0, 1, 2, 3, while on occasion we shall use Latin indices $i, j, \ldots$ for the range 1, 2, 3. Four-tensors will be denoted by capital letters ($A, B^{\mu}, C^{\mu}_{\nu}$, etc.), and three-tensors by lowercase letters (a, b_i, c_{ij}, etc.).

Poincaré transformations are linear, and so four-tensors enjoy all the special properties that follow from this: e.g. displacements Δx^{μ} are vectors, tensors at different points can be added and multiplied, and partial derivatives of tensors are tensors.

The tensor calculus is particularly rich on a space V_N endowed with a metric. This is exactly the situation for the space of events. Its metric is provided by (20.1). This can be written in tensor notation as

$$ds^2 = g_{\mu\nu}dx^{\mu}dx^{\nu}, \tag{20.3}$$

where, as in the Euclidean case, the metric tensor $g_{\mu\nu}$ is numerically constant under the permissible transformations:

$$g_{\mu\nu} = \text{diag}\,(1, -1, -1, -1). \tag{20.4}$$

Also as in that case, $g_{\mu\nu}$ is numerically equal to its conjugate $g^{\mu\nu}$:

$$g^{\mu\nu} = \text{diag}\,(1, -1, -1, -1). \tag{20.5}$$

This metric serves in the usual way [cf. (A.7)] to define the scalar product of two four-vectors **A** and **B**:

$$\mathbf{A}.\mathbf{B} = A^0 B^0 - A^1 B^1 - A^2 B^2 - A^3 B^3. \tag{20.6}$$

The space of events (x^{μ}) with metric tensor (20.4) is referred to as *Minkowski space*, or (flat) *spacetime*.

Spacetime is far more than a mere mathematical artifice. Relativity made the older concepts of absolute space and absolute time untenable, and yet they had served as a deeply ingrained framework in our minds on which to 'hang' the rest of physics. Spacetime is their modern successor: it is the new absolute framework for exact physical thought. For example, each material particle in the course of its history appears as a line (a one-dimensional aggregate of events) in spacetime: that line is called its '*worldline*'. Spacetime is filled with a tangle of such worldlines, and '*worldtubes*' for extended bodies. Our familiarity with spacetime is reinforced when we develop a 'four-tensor mentality', which is as useful in relativity as is a 'three-vector mentality' in classical physics. A four-tensor equation is automatically

Poincaré-invariant. Thus we can recognize by its form alone whether a given or proposed physical law satisfies the relativity principle, without having to work through tedious transformations. This has great heuristic value. Moreover, by automatically combining space and time, momentum and energy, electric and magnetic field, etc., the formalism illuminates some profound physical interconnections.

Care must be taken, however, not to regard spacetime as a straightforward generalization of ordinary Euclidean three-space to four dimensions, with time as just one more dimension. Owing to the distribution of signs in the metric, the time coordinate x^0 is not on the same footing as the three space-coordinates, and spacetime consequently has non-isotropic properties quite unlike Euclidean space. We discuss these and related matters in the next section.

21. The Minkowski map of spacetime

When we think of the earth, say of the configuration of its continents, we sometimes think of the real spherical earth, sometimes of its familiar Mercator map. With spacetime, most mortals, most of the time, think in terms of its 'Minkowski map'—also known as *spacetime-* or *Minkowski diagram*. This is a mapping of events onto Euclidean space $(ct, x, \ldots)$ of two, three or (theoretically sometimes) four dimensions. Our earlier Figs 3, 5, 9 are two-dimensional examples of such mappings.

Because spacetime is homogeneous (i.e. each neighbourhood is equivalent to every other), understanding one neighbourhood means understanding all. Consider therefore an arbitrary event $\mathscr{P} = (x^\mu)$ and all events $(x^\mu + \Delta x^\mu)$ in its neighbourhood. (In fact, the Δx^μ need not be small: the 'neighbourhood' could be all of spacetime.) These other events can be divided absolutely relative to $\mathscr{P}$ into three classes according to whether the square of their displacement vector from $\mathscr{P}$,

$$\Delta s^2 = c^2 \Delta t^2 - \Delta x^2 - \Delta y^2 - \Delta z^2 \tag{21.1}$$

[cf. (20.6)], is positive, zero, or negative.

The simplest class is that for which $\Delta s^2 = 0$. We have already seen in Section 6 (for differentials; the argument for Δ's is the same) that this equation characterizes events which can be connected with $\mathscr{P}$ by light signals. In fact, the equation represents a spherical light-front converging on $\mathscr{P}$ and then re-diverging from it. Alternatively, for each fixed set of ratios $\Delta x : \Delta y : \Delta z$ it represents the worldline of a photon through $\mathscr{P}$. Now consider a three-dimensional Minkowski map,

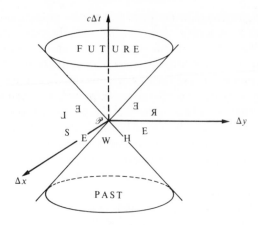

$$\text{F}\,\text{IG}.\quad 12$$

Fig. 12, of a neighbourhood of $\mathscr{P}$. It is necessarily restricted to events on one spatial plane, here an xy-plane of the inertial frame supplying the coordinates for the map. By convention, the time axis is always drawn *up*. In this map the locus of events satisfying $\Delta s^2 = 0$ is a right circular cone of semi-vertical angle $45°$, its equation being $\Delta x^2 + \Delta y^2 = (\Delta ct)^2$. Each of its 'horizontal' sections is an instantaneous map of the circular light-front centred on $\mathscr{P}$, each of its generators a photon worldline through $\mathscr{P}$. We call this locus the *light-cone* or *null cone* through $\mathscr{P}$, no matter how many dimensions (2, 3, or 4) we envisage. In full spacetime it is a three-dimensional (hyper-) surface through $\mathscr{P}$. Such cones exist at each event, and they constitute a kind of 'grain' of spacetime, which has no analogue in Euclidean space. Displacement vectors $\mathscr{P}\mathscr{Q}$ lying *on* the cone are said to be *lightlike* or *null*.

We can write (21.1) in the form

$$\Delta s^2 = \Delta t^2 (c^2 - U^2) = c^2 \Delta t^2 (1 - U^2/c^2), \qquad (21.2)$$

where $U^2 = \Sigma (\Delta x^i)^2/\Delta t^2$, from which we see how the 'signal speed' U between $\mathscr{P}$ and a neighbouring event $\mathscr{Q}$ determines the sign of Δs^2. The 'normalized' signal speed, U/c, corresponds to the slope of the displacement vector $\mathscr{P}\mathscr{Q}$ relative to the time axis in the Minkowski diagram. Thus events whose displacement vector satisfies $\Delta s^2 > 0$ lie *inside* the light-cone (i.e. where the time axis is), while those for which $\Delta s^2 < 0$ lie outside. In full spacetime the former events occur at points *within* the light-front converging to and re-diverging from $\mathscr{P}$, and the latter at points outside.

Since a particle must always move with speed less than c, its worldline must always have slope less than unity (relative to the time axis). Hence no portion of such a worldline through $\mathscr{P}$ can lie outside the cone. In particular, no particle can be present at $\mathscr{P}$ and at an event $\mathscr{S}$ outside the cone. There is, in fact, an essential space separation between $\mathscr{P}$ and $\mathscr{S}$ which, by (21.1), cannot be less than $\sqrt{(-\Delta s^2)}$. This minimum is evidently measured in a frame in which $\mathscr{P}$ and $\mathscr{S}$ are simultaneous. That such a frame exists, and that other frames exist in which the time separation between $\mathscr{P}$ and $\mathscr{S}$ is any preassigned positive or negative quantity, is seen at once by considering the set of frames on whose x-axes $\mathscr{P}$ and $\mathscr{S}$ occur, and using (7.5). Displacement vectors for which $\Delta s^2 < 0$ are said to be *spacelike*.

Similarly, it is easy to see that for any event $\mathscr{T}$ satisfying $\Delta s^2 > 0$ there exists no frame in which the time separation between $\mathscr{P}$ and $\mathscr{T}$ vanishes, that in fact it can not be less than $\Delta s/c$ in any frame, and that this minimum is measured by an observer moving uniformly between $\mathscr{P}$ and $\mathscr{T}$. On the other hand, the spatial separation between $\mathscr{P}$ and $\mathscr{T}$ can take any value. Displacement vectors for which $\Delta s^2 > 0$ are said to be *timelike*.

Timelike and lightlike displacement vectors share a property without analogue for spacelike displacement vectors: the sign of their first component is invariant. We have in fact already proved this [in Section 7(viii)] in the form that all observers agree on the time sequence—i.e. on the sign of Δt—of any two events connectible by a signal with speed $U \leqslant c$. It follows that all events on and within the upper sheet of the light-cone are regarded by *all* observers as occurring *after* $\mathscr{P}$, and they are said to constitute the *absolute future* of $\mathscr{P}$. This is also the maximal set of events that can be influenced by $\mathscr{P}$, for to get out of the light-cone a signal would have to exceed the speed of light (its slope relative to the time axis would have to exceed unity) at least *somewhere*. Similarly, events on and within the lower sheet of the light-cone constitute the *absolute past* of $\mathscr{P}$—the maximal set of events that can influence $\mathscr{P}$. Events outside the light-cone constitute the *elsewhere*: no causal relation can exist between them and $\mathscr{P}$.

All four-vectors can be represented by displacement vectors in spacetime (cf. Section A6) and therefore in the Minkowski map. Correspondingly one classifies *all* four-vectors as timelike, null (lightlike), or spacelike according to whether their square is positive, zero, or negative. And in the first two cases one further distinguishes between *future-pointing* and *past-pointing* vectors, according to the sign of the first component. But when representing vectors in the Minkowski map it is important to remember that the map distorts lengths and angles. Thus vectors that appear equally long in the diagram do not necessarily have the same 'Minkowski length' $|\mathbf{A}|$, and vice versa. For example, by rotating the hyperbolae of Fig. 5 about the t-axis we obtain hyperboloids of revolution in Fig. 12 which are loci of end-points of displacements from $\mathscr{P}$ *all* having positive or negative unit square. Also vectors that appear orthogonal in the diagram are not necessarily 'Minkowski-orthogonal' in the sense $\mathbf{A} \cdot \mathbf{B} = 0$, as exemplified by the axes of ξ and η in Fig. 5. Conversely, the axes of x' and t' *are* Minkowski-orthogonal but do not appear orthogonal in the diagram. On the other hand, it is clear that vector sums in the diagram correspond to 'Minkowski sums' $\mathbf{A} + \mathbf{B}$, that parallel vectors in the diagram correspond to 'Minkowski-parallel' vectors in the sense $\mathbf{A} = k\mathbf{B}$ for some scalar k, and that the 'Minkowski ratio' k of two such vectors is also the apparent ratio in the diagram. Consequently such theorems as the following can easily be read off from the diagram: *the sum of any number of future-pointing timelike or null vectors is itself future-pointing timelike or null, and is null if and only if all the summands are null and parallel.*

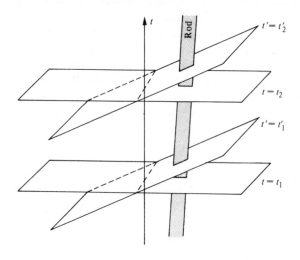

F_IG. 13

One final point: just as Americans put America in the middle of a Mercator's map, and Europeans Europe, so the frame S supplying the coordinates for a Minkowski map puts itself in the favoured position of mapping its simultaneities 'horizontally'. This is a convention without physical significance. All other frames have parallel simultaneities (instantaneous three-spaces) which—in this diagram—slant at various angles to the horizontal (see Fig. 13—and compare with the earlier Fig. 5). Consider, for example, a moving rod. Here the *absolute* object is the set of all events at the rod, i.e. its worldtube. Different observers make different sequences of world-maps of the rod by slicing its worldtube with their different (hyper-) planes t = constant.

22. Rules for the manipulation of four-tensors

(i) The rule for raising and lowering four-tensor indices follows at once from the definitions (A.10), (A.11) and from the specific values (20.4), (20.5) of the gs. Thus, for example,

$$A^0 = g^{0\mu} A_\mu = g^{00} A_0 = A_0,$$
$$A^1 = g^{1\mu} A_\mu = g^{11} A_1 = -A_1, \text{ etc.}$$

This shows that the raising or lowering of a 0 on a component leaves that component unchanged, while the raising or lowering of a 1, 2, or 3 changes the sign of the component.

(ii) Four-vectors will often be denoted by **A**, **B**, etc. [cf. before (A.7)]. Each four-vector **A** can alternatively be described by its contravariant or covariant components, A^μ or A_μ. So we could write **A** $= A^\mu$ or **A** $= A_\mu$. But we must not write $A^\mu = A_\mu$, since this would not be a true equation if interpreted numerically. So it is best to agree that when displaying components of **A** we display its contravariant components. The three contravariant spatial components A^i of a four-vector **A** are often combined into a three-vector **a**; then the following are all equivalent:

$$\mathbf{A} = A^\mu = (A^0, A^1, A^2, A^3) = (A^0, A^i) = (A^0, \mathbf{a}), \qquad (22.1)$$

$$A_\mu = (A^0, -\mathbf{a}). \qquad (22.2)$$

The reason for introducing **a** is that under a mere rotation and translation of the spatial axes in a *given* inertial frame S (which is also a Poincaré transformation!) the components A^i transform like the coordinate differences Δx^i (cf. Section A6) and those, in turn, transform like a three-vector $\mathbf{a}$ in S. For such three-vectors (and later for three-tensors) we shall use the notation $a_i = a^i$ interchangeably, i.e. the level of the indices will have no significance [cf. Exercises A(2) and A(10)]. The magnitude of three-vectors $\mathbf{a}$, $\mathbf{b}$, etc. will always be denoted by a, b, etc.

Under that subgroup of the Poincaré transformations which are homogeneous—i.e. which transform the origin-event $(0, 0, 0, 0)$ into itself and which are properly called *Lorentz* transformations—the coordinates x^μ themselves behave as a four-vector (cf. Section A6). Then we can define the four-dimensional *position vector* $\mathbf{R} = x^\mu = (ct, \mathbf{r})$ of an event relative to the origin, $\mathbf{r}$ being the three-dimensional position vector of the point at which the event occurs. For dx^μ we can always write $(c\,dt, d\mathbf{r})$.

(iii) The scalar product of two four-vectors $\mathbf{A} = (A^0, \mathbf{a})$ and $\mathbf{B} = (B^0, \mathbf{b})$ is given by

$$\mathbf{A} \cdot \mathbf{B} = g_{\mu\nu} A^\mu B^\nu = A^0 B^0 - A^1 B^1 - A^2 B^2 - A^3 B^3$$
$$= A^0 B^0 - \mathbf{a} \cdot \mathbf{b}, \tag{22.3}$$

where $\mathbf{a} \cdot \mathbf{b}$ is the usual three-vector scalar product.

(iv) When tensor transformations from one frame to another have to be actually computed, we usually find it possible to choose coordinates in standard configuration, so that the standard Lorentz transformation applies. Under it, any contravariant four-vector A^μ transforms according to the same scheme as the coordinates x^μ themselves, and this we can read off from (6.14) and (7.4) (for the inverse transformation):

$$x^{0'} = \gamma(x^0 - vx^1/c), \ x^{1'} = \gamma(x^1 - vx^0/c), \ x^{2'} = x^2, \ x^{3'} = x^3, \quad (22.4)$$
$$x^0 = \gamma(x^{0'} + vx^{1'}/c), \ x^1 = \gamma(x^{1'} + vx^{0'}/c), \ x^2 = x^{2'}, \ x^3 = x^{3'}. \quad (22.5)$$

Thus, for example, $A^{0'} = \gamma(A^0 - vA^1/c), \ldots, A^{3'} = A^3$. For the transformation of a covariant vector we need only lower the indices: $A_{0'} = \gamma(A_0 + vA_1/c), \ldots, A_{3'} = A_3$. Higher-rank four-tensors are transformed *ab initio* according to the transformation rules (A.3)–(A.5). For later reference we exhibit the required transformation coefficients $p_\mu^{\mu'}$ and $p_{\mu'}^\mu$ [cf. (A.1)], which can be read off from (22.4) and (22.5):

$$p_0^{0'} = p_1^{1'} = \gamma, \ p_0^{1'} = p_1^{0'} = -\gamma v/c, \ p_2^{2'} = p_3^{3'} = 1, \quad (22.6)$$
$$p_{0'}^0 = p_{1'}^1 = \gamma, \ p_{1'}^0 = p_{0'}^1 = \gamma v/c, \quad p_{2'}^2 = p_{3'}^3 = 1, \quad (22.7)$$

and all others vanish. Thus, for example, for a tensor $T^{\mu\nu}$,

$$T^{1'2'} = p_\mu^{1'} p_\nu^{2'} T^{\mu\nu} = p_0^{1'} p_2^{2'} T^{02} + p_1^{1'} p_2^{2'} T^{12} = \gamma (T^{12} - vT^{02}/c),$$

(22.8)

and so on.

(v) Logically we should distinguish between an inertial frame S, which is just an aggregate of defining particles; the spatial coordinate system $[x^i]$ chosen in S, which we could denote by $[S]$; and the standard coordinate system $\{x^\mu\}$ chosen in S, which also serves to coordinatize spacetime, and which we could denote by $\{S\}$. Usually S is identified with $\{S\}$, and we do not draw these distinctions too pedantically. But we *must* distinguish (for example, in the following paragraph) between the spatial axes of $[S]$ and $\{S\}$. The former are 'corpuscular'—consisting of a certain set of particles in S—and thus correspond to a plane set of worldlines in spacetime, while the latter are single lines in spacetime.

We can choose standard coordinates in spacetime in such a way that the components of a given four-vector become particularly simple. For example, if a timelike vector **A** has components $(A^0, \mathbf{a})$ in some frame S, we can reduce its components to $(A^{0'}, 0, 0, 0)$ by transforming to a frame S' that has velocity $\mathbf{v} = c\mathbf{a}/A^0$ relative to S. For the 'signal speed' along the displacement Δx^μ that represents **A** is $\Delta x^i/\Delta t = \mathbf{a}/(A^0/c)$, and so each fixed point in S' has its worldline parallel to **A**, i.e. the time axis of S' is parallel to **A**.

Similarly, if a spacelike vector **B** has components $(B^0, \mathbf{b})$ in a frame S, we can reduce its components to $(0, B^{1'}, 0, 0)$ in a suitably chosen frame S', which obviously must have its x-axis parallel to **B**. This is achieved by first aligning the *corpuscular* x-axis of S with **b**—which makes $\mathbf{B} = (B^0, B^1, 0, 0)$—and then transforming to that frame S' which is in standard configuration with S at velocity $v = cB^0/B^1$. It is easily seen from (22.4) that this indeed makes $B^{0'} = 0$.

Finally a null vector $\mathbf{C} = (C^0, \mathbf{c})$ is easily reduced to the form $(C^0, C^0, 0, 0)$ by aligning the corpuscular x-axis with **c**, but in this case we can do no more.

23.   Four-velocity and four-acceleration

Along timelike directions in spacetime $(\mathrm{d}s^2 > 0)$ it is often convenient to work with the invariant

$$\mathrm{d}\tau^2 = \frac{\mathrm{d}s^2}{c^2} = \mathrm{d}t^2 - \frac{\mathrm{d}x^2 + \mathrm{d}y^2 + \mathrm{d}z^2}{c^2}$$

(23.1)

rather than with ds^2 itself, and so we give it a special symbol and name: $d\tau$ (chosen positive in the future-pointing direction) is called the element of *proper time*. It gets its name from the fact that, for neighbouring events on a moving particle, $d\tau$ coincides with the time differential dt that is measured by an ideal clock attached to the particle, whether the particle moves uniformly or not. For in its rest frame the particle satisfies $dx = dy = dz = 0$. Alternatively, writing u for the speed of the particle, we have from (23.1)

$$\frac{d\tau^2}{dt^2} = 1 - \frac{u^2}{c^2}, \quad \frac{dt}{d\tau} = \left(1 - \frac{u^2}{c^2}\right)^{-1/2} = \gamma(u), \quad (23.2)$$

which also bears out our assertion, being just the equation for the time dilation of a moving clock. We shall not be surprised, therefore, to find the invariant $d\tau$ appearing in many relativistic formulae where in the classical analogue there is a dt.

Consider now a particle with worldline $x^\mu = x^\mu(\tau)$, τ being the proper time elapsed at the particle. (We have already seen—in Section 21—that such worldlines are entirely timelike.) Then, since $dx^{\mu'}/d\tau = p_\mu^{\mu'} dx^\mu/d\tau$, $dx^\mu/d\tau$ is a four-vector (and, in fact, a general vector). Consequently its derivative with respect to τ, $d^2x^\mu/d\tau^2$, is a four-vector also (cf. end of Section A9). These vectors are denoted by U and A, respectively, and are called the *four-velocity* and the *four-acceleration* of the particle:

$$\mathbf{U} = \frac{dx^\mu}{d\tau}, \quad \mathbf{A} = \frac{d^2x^\mu}{d\tau^2} = \frac{d\mathbf{U}}{d\tau}. \quad (23.3)$$

This process of generalizing a known three-vector by a slight modification, if necessary, of its three components and the addition of a fourth to form a four-vector, is a most fruitful way of discovering significant four-vectors and, through them, the relativistically valid laws of physics.

We obtain the relation between the four-velocity U and the three-velocity $\mathbf{u} = dx^i/dt$ by making use of (23.2):

$$\mathbf{U} = \frac{dx^\mu}{d\tau} = \frac{dx^\mu}{dt}\frac{dt}{d\tau} = \gamma(u)\frac{dx^\mu}{dt} = \gamma(u)(c, \mathbf{u}). \quad (23.4)$$

The relation between the four-acceleration A and the three-acceleration $\mathbf{a} = d^2x^i/dt^2$ is more complicated:

$$\mathbf{A} = \frac{d\mathbf{U}}{d\tau} = \gamma\frac{d\mathbf{U}}{dt} = \gamma\frac{d}{dt}(\gamma c, \gamma\mathbf{u}) = \gamma\left(\frac{d\gamma}{dt}c, \frac{d\gamma}{dt}\mathbf{u} + \gamma\mathbf{a}\right), \quad (23.5)$$

where, of course, $\gamma = \gamma(u)$. But in the instantaneous rest frame of the particle $(u = 0)$ this expression simplifies to

$$\mathbf{A} = (0, \mathbf{a}), \tag{23.6}$$

since the derivative of γ contains a factor u. Thus $\mathbf{A} = 0$ if and only if the *proper acceleration*—i.e. the magnitude of the three-acceleration in the rest frame—vanishes. The four-velocity $\mathbf{U}$, on the other hand, never vanishes.

In fact, the square of $\mathbf{U}$ is always the same, viz. c^2, no matter how the particle moves:

$$\mathbf{U}^2 = \mathbf{U} \cdot \mathbf{U} = c^2. \tag{23.7}$$

This follows at once from (23.4), most easily *after* putting $u = 0$. For $\mathbf{U}^2$, being an invariant, has the same value in all frames, so we may as well evaluate it in the rest frame. By the same artifice we find, from (23.6), that

$$\mathbf{A}^2 = -\alpha^2, \tag{23.8}$$

where α is the proper acceleration; and from (23.4) and (23.6) that

$$\mathbf{U} \cdot \mathbf{A} = 0, \tag{23.9}$$

i.e. the four-acceleration is always orthogonal to the four-velocity. [Another way to establish (23.9) is to differentiate (23.7) with respect to τ.] Equations (23.7) and (23.4) (first component positive!) show, respectively, that $\mathbf{U}$ is timelike and future-pointing; while (23.8) shows that $\mathbf{A}$ is spacelike.

$\mathbf{U}$ and $\mathbf{A}$ have analogues not only in classical kinematics, namely $\mathbf{u}$ and $\mathbf{a}$, but also in the differential geometry of curves. They are, in fact, the analogues, with respect to the particle's worldline, of the unit tangent vector dx^i/ds and principal normal vector $d^2 x^i/ds^2$ of a space curve $x^i = x^i(s)$ [cf. Exercise A(12)]. Thus α is a measure of the curvature of the worldline. (Because we have taken τ rather than s as the parameter, the actual curvature of the worldline is α/c^2.)

The four-vector calculus makes it easy to generalize our earlier formula (14.1), which expresses the proper acceleration α in terms of quantities measured in the general frame—but only in the case of rectilinear motion. From (23.5) we have, quite generally,

$$\mathbf{A}^2 = \gamma^2 \left[\dot{\gamma}^2 c^2 - (\dot{\gamma}\mathbf{u} + \gamma\mathbf{a})^2 \right], \tag{23.10}$$

where the dots denote differentiation with respect to t. Using the relation $\dot{\gamma} = \gamma^3 u\dot{u}/c^2$ [cf. (7.1)] and the familiar three-vector results $\mathbf{u}^2 = u^2$ and $\mathbf{u} \cdot \dot{\mathbf{u}} = u\dot{u}$, we find, from (23.10), the desired formula:

$$\alpha^2 = -\mathbf{A}^2 = \gamma^2 \left[\dot{\gamma}^2 u^2 + 2\gamma\dot{\gamma}u\dot{u} + \gamma^2 a^2 - \dot{\gamma}^2 c^2 \right]$$
$$= \gamma^6 u^2 \dot{u}^2/c^2 + \gamma^4 a^2. \tag{23.11}$$

In the case of rectilinear motion we have $\dot{u}^2 = a^2$, and then (23.11) reduces to (14.1), as it must. In another special case of interest, namely motion with constant speed u, (23.11) reduces to $\alpha = \gamma^2 a$. In particular, if that motion is along a circle of radius r, $a = u^2/r$ and so $\alpha = \gamma^2 u^2/r$ [cf. our earlier Exercise II(11)].

Finally we inquire into the absolute significance of the scalar product of two four-velocities, say $\mathbf{U}$ and $\mathbf{V}$, corresponding to two uniformly moving particles. Let us look at $\mathbf{U} \cdot \mathbf{V}$ in the rest frame of the second particle, in which the first has velocity $\mathbf{U} = \gamma(u)(c, \mathbf{u})$ and the second, $\mathbf{V} = (c, \mathbf{0})$. Thus

$$\mathbf{U} \cdot \mathbf{V} = c^2 \gamma(u), \tag{23.12}$$

i.e. $\mathbf{U} \cdot \mathbf{V}$ is c^2 times the Lorentz factor of the *relative* velocity of the corresponding particles. A somewhat more general formula for the scalar product of two four-vectors will be found in Exercise IV (12).

24. Wave motion

It will now be a useful exercise—both in the use of four-vectors and in the new spacetime way of thinking—to discuss the geometry of wave motion. We begin by considering a *wavefront*. This propagates some single disturbance (i.e. one of infinitesimal duration) in a frame S, and so it is a surface that progresses in time. A *fixed* surface in S has an equation of the form $f(x, y, z) = X$, where f is some function of x, y, z and X is a constant. A surface *moving* through S has a similar equation, but with time t as a parameter: $f(x, y, z, t) = X$. Different values of t correspond to different locations of the surface in time. For later convenience we shall write this equation in the standard form

$$\frac{1}{c} N(ct, x, y, z) = X, \tag{24.1}$$

where N is a differentiable function of ct, x, y, z and X is to be dimensionless. A *wavetrain*, i.e. a family of such wavefronts filling a region of space, is obtained by letting X become a continuous variable, taking one value for each moving surface. In fact, X then serves as a kind of co-moving coordinate, labelling the individual wavefronts. It—or a suitable function of it—will be called the *phase* of the wavetrain. An actual physical wavetrain corresponds to certain quantities (e.g. pressure in a sound wave, electric and magnetic field in a light-wave, etc.) being equal, or in some sense equivalent, at equal values of X. But for our present purpose of studying the purely kinematic (or geometric) aspect of wave motion, the phase is all we need consider.

There is an alternative way of looking at the wavefront (24.1) (with constant X), namely as a *hypersurface* (a three-dimensional subspace) in spacetime. Similarly the wavetrain (24.1) (with variable X) can be regarded as a one-parameter family of such hypersurfaces in spacetime.

Now, by its very form (24.1), X is a scalar, i.e. a function of position in spacetime. But the question arises whether X so defined relative to a frame S has a meaning to other observers that is independent of S. To begin with, it is clear that for a single disturbance travelling through S (i.e. a single wavefront $N/c = X$) all observers will agree on whether a given event is 'disturbed' or not. Thus, since $N/c = X$ represents the locus of disturbed events, every inertial observer obtains *his* instantaneous wavefronts by slicing the hypersurface $N/c = X$ with *his* simultaneities. We shall assume that for a wavetrain all observers agree on whether a given event coincides with a 'wavecrest', or a 'wavetrough', or whatever in between. Then all surfaces of equal phase X are indeed wavefronts (i.e. loci of crests, etc.) to *all* observers, and X has the same physical significance in all inertial frames.

The derivative $cX_{,\mu} = N_{,\mu}$ of the scalar cX is a vector (cf. Section A9) that plays an important part in the theory—especially when the waves are periodic, i.e. when they possess a well-defined frequency at each event. For convenience we shall here write N_μ for $N_{,\mu}$. Also we shall so normalize the dimensionless phase X (i.e. re-coordinatize the individual wavefronts) that X *decreases* by unity as we go from one wavecrest to the next, in the direction of propagation of the waves.

Let us, then, consider two events occurring at a *fixed* point in S, separated by a time difference Δt, and coinciding with successive wavecrests. Since the *later* wavecrest has the *bigger* phase, we shall have, by Taylor's expansion,

$$1 = \Delta X = \frac{1}{c}\left(\frac{\partial N}{\partial t}\Delta t + \frac{1}{2}\frac{\partial^2 N}{\partial t^2}\Delta t^2 + \dots\right). \qquad (24.2)$$

But now consider the so-called 'geometrical optics limit' of very high frequency v. Then the time $\Delta t = 1/v$ between successive wavecrests is very short, and we can neglect all but the first term on the right-hand side of (24.2), leaving $1 = N_0 \Delta t$. Hence $N_0 = v$.

Next consider any infinitesimal three-displacement dx^i lying entirely in an instantaneous wavefront in S. Then

$$0 = dX = \frac{1}{c} N_i dx^i,$$

i.e. N_i is normal to all such displacements dx^i [cf. (A.12)] and consequently parallel to $\mathbf{n}$, the unit normal to the wavefront in the direction of propagation. 'Raising' the index μ, we may therefore write

$$N^\mu = (v, k\mathbf{n})$$

for some three-scalar k. To determine the value of k we consider a general four-displacement $dx^\mu = (cdt, d\mathbf{r})$ from one event to another on the *same* travelling wavefront. Then $d\mathbf{r}$ is a connecting vector between two positions of the wavefront a time dt apart. For these events we have

$$0 = dX = \frac{1}{c} N_\mu dx^\mu = vdt - \frac{1}{c} k\mathbf{n} \cdot d\mathbf{r}.$$

But clearly $\mathbf{n} \cdot d\mathbf{r} = wdt$, where w is the (phase-) velocity of the wave. It follows that $k = vc/w$, and so we have found

$$\mathbf{N} = N^\mu = v\left(1, \frac{c}{w}\mathbf{n}\right). \tag{24.3}$$

For obvious reasons $\mathbf{N}$ is called the *frequency four-vector* of the wave (or sometimes simply the *wave vector*). It is significant that the three kinematic characteristics of a wave at a given event—v, w, and $\mathbf{n}$—combine in this particular way to form a four-vector, and this fully determines their transformation properties from one inertial frame to another (as we shall demonstrate below).

Given a wavetrain (24.1), we define its *four-rays* as those curves in spacetime which are everyhere in the direction of $\mathbf{N}$. Hence they satisfy, for some parameter θ, the equation

$$\frac{dx^\mu}{d\theta} = N^\mu. \tag{24.4}$$

Evidently these curves are the orthogonal trajectories of the hypersurfaces $N/c = X$ representing the wavetrain (since $N_\mu \mathrm{d}x^\mu = 0$ for displacements $\mathrm{d}x^\mu$ *in* each hypersurface $N = $ constant). Similarly the *three-rays* $\mathrm{d}x^i/\mathrm{d}\theta = N^i$—the 'projections' of the four-rays into an instantaneous three-space $t = $ constant of some inertial frame S—are the orthogonal trajectories of the instantaneous wavetrain in S. But regarded as *worldlines*, what particles do the four-rays represent? Can we think of them as particles that ride with the wavefronts along the three-rays? In general, no! For if $\mathrm{d}x^\mu = (c\mathrm{d}t, \mathrm{d}\mathbf{r}) \propto \left(1, \dfrac{c}{w}\mathbf{n}\right)$, we evidently have, for the velocity $\mathbf{v}$ of these particles in S,

$$\mathbf{v} = \frac{\mathrm{d}\mathbf{r}}{\mathrm{d}t} = \frac{c^2}{w}\mathbf{n}. \tag{24.5}$$

So if the wave velocity w exceeds c, the particle velocity v does not, and vice versa. Only in the case of *light-waves* do these 'particles' indeed ride with the waves: they will turn out to be photons. In all other cases the waves are the so-called *de Broglie waves* associated with the particles whose worldlines are the four-rays.

  In the case of light-waves in vacuum, $w = c$ everywhere, and so, from (24.3),

$$\mathbf{N} = v(1, \mathbf{n}), \tag{24.6}$$

which is evidently *null*: $g_{\mu\nu}N^\mu N^\nu = 0$. Differentiating this equation with respect to x^σ yields $2g_{\mu\nu}N^\mu N^\nu_{,\sigma} = 0$, i.e. [cf. Exercise A(11)]

$$N^\mu N_{\mu,\sigma} = 0. \tag{24.7}$$

But also, since partial derivatives commute, we have

$$N_{\mu,\nu} = N_{\nu,\mu}. \tag{24.8}$$

'Transvecting' this equation with N^ν (one uses the term *transvect* rather than *multiply* when an *inner* product is formed), we then find, by reference to (24.7),

$$N_{\mu,\nu}N^\nu = 0. \tag{24.9}$$

Lastly let us differentiate the equation of the rays, (24.4), and then successively use both it and (24.9) to obtain

$$\frac{\mathrm{d}^2 x^\mu}{\mathrm{d}\theta^2} = N^\mu_{,\nu}\frac{\mathrm{d}x^\nu}{\mathrm{d}\theta} = N^\mu_{,\nu}N^\nu = 0. \tag{24.10}$$

This equation tells us that light-rays are *straight* (both in three and four dimensions)—a beautiful result, which enables us to identify light-rays with photons. It follows simply from the speed of the waves being c, and is independent of their shape and other properties.

Light-rays have another important characteristic. Like all rays they are the orthogonal trajectories of the wavetrain in spacetime, but unlike other rays they do not *cut* the wavetrain hypersurfaces: they lie entirely *within* them. The reason is that in the case of light these hypersurfaces are *null* (i.e. have a null normal N^μ) so that a displacement $dx^\mu \propto N^\mu$ satisfies the condition $N_\mu dx^\mu = 0$ for being *in* the surface. Figure 14 illustrates this: it shows the light-cones (wavefronts) emitted by a point source travelling along a worldline l; the four-rays are the generators of the cones.

An important special case of wave motion, and in fact the simplest, is that of *plane waves*, moving with constant v, w, and $\mathbf{n}$. We then have

$$N_\mu = v\left(1, \ -\frac{c}{w}\mathbf{n}\right) = \text{constant}, \qquad (24.11)$$

whence the basic equation $X_{,\mu} = \frac{1}{c}N_\mu$ can be integrated at once to

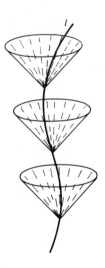

F<small>IG</small>.   14

give

$$X = \frac{1}{c} N_\mu x^\mu = v\left(t - \frac{\mathbf{r} \cdot \mathbf{n}}{w}\right), \qquad (24.12)$$

from which we have omitted, without loss of generality, the constant of integration. For the discussion of plane waves it is well to restrict oneself to inertial frames related by *homogeneous* Lorentz transformations, so that the x^μ occurring in (24.12) is a four-vector. (Otherwise the freedom of adding a constant of integration can be used to replace x^μ by the coordinate increments Δx^μ from some arbitrary event.) Typical tensor disturbances propagated with plane waves are of the form

$$A_\nu^{\mu\,\cdots} = \overset{(0)}{A_\nu^{\mu\,\cdots}} \cos 2\pi X = \overset{(0)}{A_\nu^{\mu\,\cdots}} \cos 2\pi v \left(t - \frac{\mathbf{r} \cdot \mathbf{n}}{w}\right), \quad (24.13)$$

where $\overset{(0)}{A_\nu^{\mu\,\cdots}}$ is the tensor amplitude of the wave. It should be noted that, because of (24.11), we need not in the case of plane waves restrict ourselves to the high-frequency limit in order for our earlier analysis to be valid: in the expansion (24.2) the quadratic and higher terms are now *automatically* zero.

Our final use for the moment of the main result (24.3) of the present section will be to derive the transformation properties of periodic wavetrains. In particular, this can then replace our earlier 'naive' derivations of the Doppler and aberration formulae for light waves given in Chapter III. Consider, therefore, two inertial frames S and S′ in standard configuration. In S let a wavetrain at the origin, say, travel with velocity w and frequency v in a direction $\mathbf{n} = -(\cos\alpha, \sin\alpha, 0)$ making an angle α *with the negative x-axis* (as in Sections 17 and 18). Then, by (24.3),

$$N^\mu = \left(v, -\frac{cv}{w}\cos\alpha, -\frac{cv}{w}\sin\alpha, 0\right). \qquad (24.14)$$

Transforming these components to S′ by the standard transformation scheme (22.4), we find [with $\gamma = \gamma(v)$]

$$v' = v\gamma\left(1 + \frac{v}{w}\cos\alpha\right), \qquad (24.15)$$

$$\frac{v'\cos\alpha'}{w'} = \frac{v\gamma(\cos\alpha + vw/c^2)}{w}, \qquad (24.16)$$

$$\frac{v'\sin\alpha'}{w'} = \frac{v\sin\alpha}{w}. \qquad (24.17)$$

The first of these equations expresses the Doppler effect for waves of all phase velocities and, in particular, for light-waves ($w = c$). In the latter case it is seen to be equivalent to our previous formula (17.3).

From (24.16) and (24.17) we obtain the general wave aberration formula

$$\tan \alpha' = \frac{\sin \alpha}{\gamma \left(\cos \alpha + vw/c^2\right)}. \tag{24.18}$$

In the particular case when $w = c$ this is seen to be equivalent to our previous equations (18.1) and (18.2)–in fact, it corresponds to their quotient.

To get the transformation of w, we *could* eliminate the irrelevant quantities from (24.15)–(24.17), but it is simpler to make use of the invariance of $\mathbf{N}^2$. Writing this out in S and S', we have, from (24.3),

$$v^2 \left(1 - \frac{c^2}{w^2}\right) = v'^2 \left(1 - \frac{c^2}{w'^2}\right), \tag{24.19}$$

whence, by use of (24.15),

$$1 - \frac{c^2}{w'^2} = \frac{(1 - c^2/w^2)(1 - v^2/c^2)}{[1 + (v/w)\cos \alpha]^2}. \tag{24.20}$$

This formula does not, in general, transform speeds in the same way as (13.5). The reason is that a particle riding the crest of a wave in the direction of the wave normal in one frame does not, in general, do so in another frame: there it rides the crest of the wave also, but not in the normal direction [cf. Exercise III(14)]. The exceptional cases are (i) when $w = c$, and, of course, (ii) when S and S' move in the same direction as the wave. Nevertheless, like (13.5), (24.20) transforms $w \gtrless c$ into $w' \gtrless c$, respectively.

Exercises IV

1. Theorem: *A transformation which transforms a metric $g_{\mu\nu}dx^\mu dx^\nu$ with constant coefficients into a metric $g_{\mu'\nu'}dx^{\mu'}dx^{\nu'}$ with constant coefficients must be linear.* Fill in the details of the following proof: $g_{\mu\nu}dx^\mu dx^\nu = g_{\mu'\nu'}dx^{\mu'}dx^{\nu'} = g_{\mu'\nu'}p_\mu^{\mu'}p_\nu^{\nu'}dx^\mu dx^\nu$, so $g_{\mu\nu} = g_{\mu'\nu'}p_\mu^{\mu'}p_\nu^{\nu'}$. Differentiating with respect to x^σ we get $g_{\mu'\nu'}p_{\mu\sigma}^{\mu'}p_\nu^{\nu'} + g_{\mu'\nu'}p_\mu^{\mu'}p_{\nu\sigma}^{\nu'} = 0$ (i). Interchange μ and σ to form equation (ii). Again, in (i) interchange ν and σ to form (iii). Subtract (iii) from the sum of (i) and (ii). Then $2g_{\mu'\nu'}p_{\mu\sigma}^{\mu'}p_\nu^{\nu'} = 0$. Transvect first by $p_{\sigma'}^\nu$ and then by $g^{\nu'\sigma'}$ and obtain $p_{\mu\sigma}^{\nu'} = 0$. This proves linearity.

2. Prove that, for any straight-line segment in spacetime, $|\Delta s|$ $= \int |ds|$, where Δs^2 is defined by (21.1) and ds^2 is similarly defined in terms of the differentials. This result, of course, is necessary for the consistency of the notation. [*Hint*: let the segment have equation $x^\mu = A^\mu \theta + B^\mu$, $0 \leqslant \theta \leqslant 1$.]

3. All vectors in this problem are presumed to be real and non-zero. Let T, S, N, V, respectively denote timelike, spacelike, null, and general vectors. Prove: (i) any V orthogonal to a T or N (other than the N itself) is an S; (ii) the sum of two Ts, or of a T and an N, which are *isochronous* (i.e. both pointing into the future or both into the past) is a T isochronous with them; (iii) the sum [difference] of two isochronous Ns is a $T[S]$, or, in the case of two parallel Ns, an N; (iv) every $T[S]$ is expressible as the sum [difference] of two isochronous Ns; (v) the scalar product of two Ts, or of a T and an N, which are isochronous, is positive; that of two isochronous Ns is positive, unless they are parallel, in which case it is zero. [*Hint*: the component specializations of Section 22(v) may help; so may a spacetime diagram, to organize ideas.]

4. The aggregate of events considered by an inertial observer to be simultaneous at his time $t = t_0$ is said to be the observer's instantaneous three-space $t = t_0$. Show that the join of any two events in such a space is orthogonal to the observer's worldline, and that, conversely, any two events whose join is orthogonal to the observer's worldline are considered simultaneous by him.

5. (i) Prove the *zero-component lemma* for four-vectors: if a four-vector V^μ has a particular one of its four components zero in *all* inertial frames then the entire vector must be zero. [*Hint*: suppose first $V^1 \equiv 0$; if there is a frame in which $V^2 \neq 0$ or $V^3 \neq 0$, we can rotate the spatial axes to make $V^1 \neq 0$; if there is a frame in which $V^0 \neq 0$, we can apply a Lorentz transformation to make $V^1 \neq 0$; and so on.] (ii) If $A_{\mu\nu}$ is a *symmetric* tensor and $A_{00} \equiv 0$, prove $A_{\mu\nu} = 0$. Is this result true for the vanishing of *any* one component of $A_{\mu\nu}$?

6. An antisymmetric tensor $T^{\mu\nu}$ has the following components in a frame S:

$$T^{\mu\nu} = \begin{pmatrix} 0 & -e_1 & -e_2 & -e_3 \\ e_1 & 0 & -b_3 & b_2 \\ e_2 & b_3 & 0 & -b_1 \\ e_3 & -b_2 & b_1 & 0 \end{pmatrix}.$$

(i) Find the values of all the components $T^{\mu'\nu'}$ in the usual second frame S'. (ii) Verify directly that $\frac{1}{2}T_{\mu\nu}T^{\mu\nu} = (b_1^2 + b_2^2 + b_3^2) - (e_1^2 + e_2^2 + e_3^2)$ is an invariant—which is of course implicit in its tensor form. (iii) Exhibit $\overset{*}{T}_{\mu\nu}$ in component form, and verify $\overset{*}{T}_{\mu\nu}\overset{*}{T}^{\mu\nu} = -T_{\mu\nu}T^{\mu\nu}$. [Cf. Exercises A(17), (18).]

7. Consider Euclidean three-space and in it the group of transformations consisting of rotations and translations of the right-handed orthogonal axes of x^i. These we shall call RT transformations, and they are clearly linear. Three-tensors are objects that behave as tensors under this transformation group [cf. Exercises A(2), A(10); translations have no effect on tensor components]. Evidently RT transformations (augmentud by $t' = t$) form a subgroup of the Poincaré group. Prove: (i) if an RT transformation is applied to a four-vector $A^\mu = (A^0, \mathbf{a})$, A^0 transforms as a scalar and $\mathbf{a}$ as a three-vector; (ii) if an RT transformation is applied to a four-tensor $T^{\mu\nu}$, then T^{00} transforms as a scalar, T^{0i} and T^{i0} transform as three-vectors, and T^{ij} transforms as a three-tensor. Note that corresponding to any three-vector b^i we can define a 'dual' tensor $b_{ij} = \varepsilon_{ijk}b^k$, so that $(b^1, b^2, b^3) = (b_{23}, b_{31}, b_{12})$. Accordingly, the bs in the preceding exercise transform as a three-vector under RT transformations, as the notation suggests.

8. An inertial observer O has four-velocity $\mathbf{U}_0$ and a particle P has (variable) four-acceleration $\mathbf{A}$. If $\mathbf{U}_0 \cdot \mathbf{A} = 0$, what can you conclude about the speed of P in O's rest frame?

9. Use the fact that $\mathbf{U} = \gamma(u)(c, \mathbf{u})$ is a four-vector to rederive the transformation equations (13.3) and (13.6) for $\mathbf{u}$. [Cf. after (22.5).].

10. Use the fact that $\mathbf{A}$ as given in (23.5) is a four-vector to give an alternative solution to the first part of Exercise II(11).

11. We shall say that three particles move *codirectionally* if their three-velocities are parallel in some inertial frame. Prove that the necessary and sufficient condition for this to be the case is that the four-velocities $\mathbf{U}, \mathbf{V}, \mathbf{W}$ of these particles be linearly dependent. [*Hint*: the component specializations of Section 22(v) may help.]

12. For any two future-pointing timelike vectors $\mathbf{V}_1$ and $\mathbf{V}_2$, prove that $\mathbf{V}_1 \cdot \mathbf{V}_2 = V_1 V_2 \cosh \phi_{12}$, where ϕ_{12}, the 'hyperbolic angle' between $\mathbf{V}_1$ and $\mathbf{V}_2$, equals the relative rapidity of two particles having $\mathbf{V}_1$ and $\mathbf{V}_2$ as worldlines. [*Hint*: (23.12).] Moreover, prove that ϕ is additive, i.e. for any three *coplanar* vectors $\mathbf{V}_1, \mathbf{V}_2, \mathbf{V}_3$, (corresponding to codirectional particles), $\phi_{13} = \phi_{12} + \phi_{23}$. For certain spacelike vectors $\mathbf{W}_1$ and $\mathbf{W}_2$ we can write $\mathbf{W}_1 \cdot \mathbf{W}_2 = -W_1 W_2 \cos\theta_{12}$; when is that possible, what is then the meaning

of θ_{12}? Note from the above that for two timelike vectors $|\mathbf{U} \cdot \mathbf{V}| \geqslant UV$, while for *such* spacelike vectors $|\mathbf{U} \cdot \mathbf{V}| \leqslant UV$.

13. In a given inertial frame S_0 moving with four-velocity U_0^μ, a tensor $T^{\mu\nu}$ has but a single non-vanishing component: $T^{00} = c^2$. Find the components of this tensor in the general frame S, relative to which $U_0^\mu = \gamma(u)(c, \mathbf{u})$. [*Hint*: once you see the trick, this is a very easy problem.]

14. A particle moves rectilinearly with constant proper acceleration α. If $\mathbf{U}$ and $\mathbf{A}$ are its four-velocity and four-acceleration, τ its proper time, and *units are chosen to make $c = 1$*, prove that $(d/d\tau)\mathbf{A} = \alpha^2 \mathbf{U}$. [*Hint*: Exercise II(14).] Prove, conversely, that this equation, *without* the information that α is the proper acceleration, or constant, implies both these facts. [*Hint*: differentiate the equation $\mathbf{A} \cdot \mathbf{U} = 0$ and show that $\alpha^2 = -\mathbf{A} \cdot \mathbf{A}$.] And finally show, by integration, that the equation implies rectilinear motion in a suitable inertial frame, and thus, in fact, hyperbolic motion. Consequently $(d/d\tau)\mathbf{A} = \alpha^2 \mathbf{U}$ is the tensor equation characteristic of hyperbolic motion.

15. By use of (23.4) and (23.5) show that the tensor equation of the preceding exercise is exactly equivalent to the three-vector equation

$$\mathbf{b} + 3\gamma^2(\mathbf{u} \cdot \mathbf{a})\mathbf{a} = 0, \quad \text{i.e.} \quad \frac{d}{dt}(\gamma^3 \mathbf{a}) = 0,$$

where $\mathbf{a} = d\mathbf{u}/dt$ and $\mathbf{b} = d\mathbf{a}/dt$, which therefore holds in all frames if it holds in one. Deduce that hyperbolic motion is fully characterized by the property that in each instantaneous rest frame $\mathbf{b} = 0$. [*Hint*: recall that $\dot\gamma = \gamma^3 \mathbf{u} \cdot \mathbf{a}$—cf. after (23.10).]

16. Rederive the frequency transformation formula (24.15) by using the invariance of $\mathbf{N} \cdot \mathbf{U}$, where $\mathbf{U}$ is the four-velocity of the frame S'.

17. (i) If $w > c$, prove that the relation between a four-ray and a wavefront is locally the same as that between an observer's worldline and one of his instantaneous three-spaces. (ii) Since v is not an invariant (it can be arbitrarily large or small in different frames) what is the logic of the geometric optics limit of very large v?

18. Prove the statement at the end of Section 24 to the effect that the respective speed transformation formulae (24.20) and (13.5) for waves and particles coincide when (i) $w = c$ or (ii) the frames S and S' move in the same direction as the wave.

V

RELATIVISTIC PARTICLE MECHANICS

25. Introduction

Our studies so far have essentially been elaborations of Einstein's two basic axioms, without the addition of further hypotheses. Thus we developed the spacetime structure, and some kinematic properties of light-propagation, implicit in those axioms. But now we come to the next point in the programme of special relativity, namely a scrutiny of the existing laws of physics and the modification of those that are found not to be Lorentz-invariant. At that stage further hypotheses are needed. For there is no logical or empirical way to *prove* a basic law of physics. It is a mathematical model—a human invention—that must be consistent with our limited experience, but it must also allow us to make new predictions that could, at least in principle, be falsified by further experiments. Such predictions and the attempts to falsify them (usually described as attempts to *verify* them) are in themselves a main stimulus to the progress of physics. Laws that cannot be falsified are tautologies. Yet Nature is strangely simple. Within a suitable mathematical framework, some of the simplest imaginable laws are the ones that Nature seems to follow, at least to the accuracy that we are able to test.

Our first target will be Newton's particle mechanics (not including, however, his theory of gravity, which is dealt with in Einstein's *general* relativity). Since it is not Lorentz-invariant, it must be modified. Among its obvious theoretical defects—in the light of special relativity—are that it allows particles to be accelerated to arbitrarily large speeds, and that it asserts the equality of action and reaction of distantly interacting particles at any instant, but whose instant? Still, it is noteworthy that Newtonian *particle* mechanics was modified long before any of its deficiencies had shown up empirically. The new mechanics is known as 'relativistic' mechanics. This is not really a good name, since, as we have seen, Newton's mechanics is relativistic too, but under the 'wrong' (Galilean) transformation group. Newton's theory has excellently served astronomy (e.g. in foretelling eclipses and orbital motions in general), it has been used as the basic theory in the incredibly delicate operations of sending probes to the Moon and some of the planets, and it has proved itself reliable in countless terrestrial applications. Thus it cannot be *entirely* wrong. Before the twentieth century, in fact, only a single case of irreducible failure was known, namely the excessive advance of the perihelion of the planet Mercury, by about 43 seconds (!) of arc per century. Since the advent of modern particle accelerators, however, vast discrepancies with Newton's laws have been uncovered, whereas the new mechanics consistently gives correct descriptions. (Of course, mechanics has undergone *two* 'corrections', one due to relativity and one due to quantum theory. We are here concerned exclusively with the former.) The new mechanics practically overlaps with the old in a large domain of applications (dealing with motions that are slow compared to the speed of light) and, in fact, it delineates the domain of sufficient validity of the old mechanics as a function of the desired accuracy. Roughly speaking, the old mechanics is in error to the extent that the γ-factors of the various motions involved exceed unity. In laboratory collisions of elementary particles γ-factors of the order of 10^4 are reached, and γ-factors as high as 10^{11} have been calculated for some cosmic-ray protons incident in the upper atmosphere. Applied to such situations, Newtonian mechanics is not just *slightly* wrong: it is totally wrong. Yet within its known slow-motion domain, Newton's theory will undoubtedly continue to be used for reasons of conceptual and technical convenience. And as a logical construct it will remain as perfect and inviolate as Euclid's geometry. Only as a model of nature it must not be stretched unduly.

26. The conservation of four-momentum

There are many ways to approach the new mechanics of particle collisions and of particles in external fields, and all lead to the same, by now, well-proven theory. Historically the first approaches were via Maxwell's theory, and in particular via the Lorentz force as the prototype of a relativistic force. But this gives undue logical prominence to electromagnetic theory, and, in fact, to the force concept, although that was primary also in Newton's scheme. Collision mechanics, however, can go a long way on a 'lower' level, without the need to define force. Analogues of the laws of conservation of energy and momentum, which are *derived* results in Newton's theory, can here usefully be taken as primary. But these, again, can be arrived at in different ways—either formally or by various physical arguments. We here choose the formal approach, as being certainly the quickest, and perhaps also the most honest.

Imagine special relativity had been discovered only last year, and, by enormous effort, had been brought to the stage that we have reached at the end of the last chapter (which already includes acceptance of Newton's first law). It is now up to us to 'invent' relativistic mechanics, but at first only collision mechanics, since that is what is most needed by the experimenters. How would we *really* set about doing this—before we publish and cover our tracks? Surely we would doodle with four-vectors, in the hope of finding some useful four-vector law (simpler than a four-*tensor* law) which, as we have seen, would automatically be Lorentz-invariant. We don't know too many four-vectors, but we would expect the velocity and acceleration vectors, **U** and **A**, of the various particles to play a role. By analogy with Newtonian theory, we might assume that the mass m of a particle is a scalar, and then define a four-momentum $\mathbf{P} = m\mathbf{U}$ and a four-force $\mathbf{F} = m\mathbf{A}$. A little more doodling, however, convinces us that **A** is really irrelevant in collisions, and so we put **F** aside for later consideration. For the moment, we try a law that asserts the *conservation of four-momentum*, i.e. that the sum of the four-momenta of all the particles going into a collison is the same as the sum of the four-momenta of all those coming out. (The collision may or may not be elastic, and there may be more, or fewer, or other particles coming out of it than going in.) We can write this in the form

$$\sum{}^{*} \mathbf{P}_{(a)} = 0, \qquad (26.1)$$

where $a = 1, 2, \ldots$ (*not* a tensor index!) refers to the various particles both before and after the collision, and Σ^* is a sum that counts pre-collision terms positively and post-collision terms negatively. This will turn out to be a very satisfactory law from all points of view, both empirical and theoretical.

For reasons that will soon become apparent, we shall write m_0 for our original scalar mass m. Then, by (23.4), the four-momentum **P** of an arbitrarily moving particle is given by

$$\mathbf{P} = m_0 \mathbf{U} = m_0 \gamma(u)(c, \mathbf{u}) = (mc, \mathbf{p}), \qquad (26.2)$$

where, in the last equation we have written

$$m = \gamma(u)m_0, \qquad (26.3)$$
$$\mathbf{p} = m\mathbf{u}. \qquad (26.4)$$

The formalism leads us naturally to this quantity m, which we shall call *relativistic inertial mass* (or often just 'mass'), and to $\mathbf{p}$, which we shall call *relativistic momentum* (or often just 'momentum'). Whatever its physical significance may be, observe that m increases with speed; when $u = 0$ it is least, namely m_0, which we call the *rest mass* of the particle.

Now if the four-vector $\Sigma^*\mathbf{P}$ vanishes [we shall leave off the as in (26.1) for brevity], each of its components must vanish separately and so, in particular, we must have $\Sigma^*\mathbf{p} = 0$, i.e.

$$\sum{}^* m\mathbf{u} = 0. \tag{26.5}$$

This *looks* exactly like Newton's law of momentum conservation; but now each mass m has a velocity-dependence (26.3). Is that reasonable? Reasonable or not, it is the only possible Lorentz-invariant law of the form (26.5). For suppose there were another such law, say

$$\sum{}^* \tilde{m}\mathbf{u} = 0 \tag{26.6}$$

(e.g. with velocity-independent $\tilde{m}$s, or with $\tilde{m}$s that are other functions of u). Then consider, in some frame S_0, a collision of two *identical* particles A and B, approaching each other with equal and opposite velocities $\pm\mathbf{u}$, and fusing upon impact. By symmetry, the resultant compound particle C must be at rest, no matter what specific law (26.6) is adopted. Now transform this experiment to an arbitrary frame S. There is, of course, only one way to transform the three-velocities of A, B, C, and these will uniquely determine the ratio of the $\tilde{m}$s in S via (26.6). But a possible set of $\tilde{m}$s is given by the Lorentz-invariant law (26.3), whence, in particular,

$$\frac{\tilde{m}_A}{\tilde{m}_B} = \frac{m_A}{m_B} = \frac{\gamma(u_A)}{\gamma(u_B)},$$

in an obvious notation, and so

$$\frac{\tilde{m}_A}{\gamma(u_A)} = \frac{\tilde{m}_B}{\gamma(u_B)}. \tag{26.7}$$

But on what could the common value of these ratios depend in different frames? Clearly it must be constant, which shows that the $\tilde{m}$s, too, are of the form (26.3). Alternatively, consider a sequence of frames S such that $u_B \to 0$. Then we find from (26.7) that, in the limit, $\tilde{m}_A = \gamma(u_A)(\tilde{m}_B)_0 = \gamma(u_A)(\tilde{m}_A)_0$, where we have written $(\tilde{m}_A)_0 = (\tilde{m}_B)_0$ for the 'masses at rest' of the identical particles A, B. Since in such experiments u_A can have any value, our assertion is established.

Thus we see that if we wish to salvage Newton's law of momentum conservation in the form (26.5), we must allow m to vary precisely as in (26.3). Moreover, (26.5) reduces to the Newtonian law itself (with velocity-independent ms) when all the relevant speeds are small. But for such situations Newton's law is extremely well validated. So if we look at a slow-motion (Newtonian) collision taking place in a frame S_0, from another frame S at high velocity relative to S_0, the same collision automatically satisfies the new law in S, by its Lorentz invariance. This is surely another recommendation for it, and the reader may by now be ready to accept, on purely theoretical grounds, the new law we have 'discovered'. Needless to say, it is in fact universally accepted as the basis of relativistic mechanics, and has been thoroughly validated by observations of elementary particle collisions for the last sixty years.

The rest mass m_0 is an *invariant*, i.e. all observers agree on its value at any instant of a particle's history. But we have no guarantee that it is *constant*: the rest mass of a particle may be altered in a collision if its internal state changes, and it may also be altered by passage through certain fields of force. But for a *freely* moving particle (even if it undergoes internal changes of state), m_0 must be constant, by Newton's first law and momentum conservation. It is equivalent to the 'Newtonian' mass of the particle, i.e. that which is determined in slow-motion collisions.

27. The equivalence of mass and energy

Let us now return to our deceptively simple looking equation (26.1). We have yet to extract its most profound implication. Just as the conservation of the four-momentum **P** implies (26.5)—the conservation of its spatial part—so it also implies the conservation of its temporal part, the mass:

$$\sum{}^* m = 0 \qquad (27.1)$$

[cf. (26.2)]. At first sight, this appears to be just an analogue of the Newtonian law of mass conservation, and as such trivial and easy to accept. But the Newtonian law expressed a belief in the permanence of matter, which is now known to be false: matter can be 'destroyed' by being converted into radiation. So it is just as well that (27.1) is *not*, in fact, an analogue of the Newtonian law (except in a purely formal sense). It asserts the conservation not of matter but of a quantity, m, that varies with speed. Classically we know of only one such quantity, namely the kinetic energy of particles in an *elastic* collision. Of course, (27.1) holds in *all* collisions—not just elastic ones—if (26.1) does. Could it be that m (or a multiple of it) is some generalization of kinetic energy—some *total* energy—that is conserved in all collisions? The answer turns out to be 'yes' and was regarded by Einstein, who found it, as the most significant result of his special theory of relativity. Nevertheless Einstein's assertion of the full equivalence of mass and energy, according to the famous formula

$$E = mc^2, \tag{27.2}$$

is in part a *hypothesis*, as we shall see. It cannot be uniquely deduced from the basic law (26.1).

Consider the following expansion for the mass (26.3):

$$m = m_0 \left(1 - \frac{u^2}{c^2}\right)^{-1/2} = m_0 + \frac{1}{c^2}\left(\tfrac{1}{2}mu^2\right) + \ldots \tag{27.3}$$

This shows that the relativistic mass of a slowly moving particle exceeds its rest-mass by $1/c^2$ times its kinetic energy (assuming the approximate validity of the Newtonian expresssion for the latter). So kinetic energy *contributes* to the mass in a way that is consistent with (27.2). In fact, it is equation (27.3) that supplies the constant of proportionality between E and m.

We can next generalize from the mass contribution of kinetic energy to that of all forms of energy. For, the second basic property that we associate with energy, after its conservation, is its transmutability from one form into another. Suppose, for example, that two particles at room temperature collide inelastically to form a doublet at rest, which then gives up ΔE units of heat energy in cooling to room temperature. By energy conservation, ΔE equals the original kinetic energy of the two particles; but then, by mass conservation, the doublet just after impact had a mass $\Delta E/c^2$ units in excess of the rest-masses of the two original particles and thus in excess of its own rest-mass at room temperature. So heat energy also contributes to mass according to Einstein's formula. But then *all* energy must so contribute. For any kind of internal energy that a particle might possess (i.e. energy other than kinetic energy) can be transmuted into heat without affecting the particle's velocity or rest-mass (see end of Section 26).

Yet it is still logically possible that energy only *contributes* to mass, without causing *all* of it. Especially in Einstein's time it would have been perfectly reasonable to suppose that the elementary particles are indestructible, so that the *available* energy of a macroscopic particle would be $c^2(m-q)$, where q is the total rest-mass of its constituent elementary particles. In collisions $\Sigma c^2 m$, $\Sigma c^2 q$, and $\Sigma c^2(m-q)$ would then all be conserved, but only the last could properly be called energy. To equate *all* mass with energy required an act of aesthetic faith, very characteristic of Einstein. Of course today we know how amply Nature has confirmed that faith. Not only have we learned (in 1934) of 'pair annihilation' and 'pair creation' in which an elementary particle and its antiparticle annihilate each other and set free a corresponding amount of radiative energy, or vice versa; but also of the spontaneous decay of neutral mesons into photon pairs; and of collisions in which different elementary particles emerge than went in. Moreover, Einstein's mass–energy equivalence is not restricted to collision mechanics. It has been found applicable and valid in many branches of physics, from electromagnetism to general relativity. It is truly a new fundamental principle of physics.

We shall distinguish between the *kinetic* energy T, which a particle possesses by virtue of its motion,

$$T = c^2(m - m_0), \tag{27.4}$$

and its *internal* energy $c^2 m_0$. Note that $T \to \infty$ as $u \to c$, which shows how nature prevents violations of the relativistic speed limit. For an 'ordinary' particle the internal energy is vast: in each gram of mass there are 9×10^{20} ergs of energy, roughly the energy of the Hiroshima bomb (20 kilotons). A very small part of this energy resides in the thermal motions of the molecules constituting the particle, and can be given up as heat; a part resides in the intermolecular and interatomic cohesion forces, and some of that can be given up in chemical explosions; another part may reside in excited atoms and escape in the form of radiation; much more resides in nuclear bonds and can also sometimes be set free, as in the atomic bomb. But by far the largest part of the energy (about 99 per cent) resides simply in the mass of the ultimate particles, and cannot be further explained. Nevertheless, it too can be liberated under suitable conditions, e.g. when matter and antimatter annihilate each other.

If every form of energy has mass, we would expect light to have mass and thus to bend in a gravitational field like that of the sun.

Indeed, this has been observed.[1] The radiation which the sun itself pours into space represents a mass loss to it of more than four million tons per second! Radiation, having mass and velocity, must also have momentum; accordingly, the radiation from the sun is a (small) contributing factor to the observed deflection of the tails of comets away from the sun. (The major factor is the corpuscular 'solar wind'.) The mass–energy equivalence is also demonstrated on the atomic scale. Thus the total mass of the separate components of a stable atomic nucleus always exceeds the mass of the nucleus itself, since energy (i.e. mass) would have to be supplied to decompose the nucleus against the nuclear binding forces. This is the reason for the well known 'mass defect'. Nevertheless, if a nucleus is split into two new nuclei, these parts may have greater *or* lesser mass than the whole. (With the lighter atoms, the parts usually exceed the whole, whereas with the heavier atoms the whole can exceed the parts owing to various internal rearrangements.) In the first case, energy can be released by 'fusion', in the second, by 'fission'.

There are two final important theoretical points to be noted. First, by arguments similar to, but somewhat more complicated than those we gave for the unicity of a Lorentz-invariant law of the form $\Sigma^* m\mathbf{u} = 0$, it can be shown[2] that the only velocity-dependent quantity m that can enter a Lorentz-invariant conservation law of the form $\Sigma^* m = 0$ is a linear function of $\gamma(u)$, such as $\gamma(u)m_0$ + constant. So if there exists a conserved energy at all, we have found it.

The second point is that the conservation laws $\Sigma^* m = 0$ and $\Sigma^* m\mathbf{u} = 0$ are not independent. In fact, we cannot have the one without the other. This is a direct consequence of the 'zero component' lemma [cf. Exercise IV(5)], according to which the universal vanishing of one component of a four-vector—in our case $\Sigma^* \mathbf{P}$—implies the vanishing of all the others.

---

[1] The exact amount of bending is affected by the curvature of spacetime around the sun, and is predicted by general relativity. Its verification was therefore crucial for the acceptance of that theory. This provided the impetus for the measurements, first performed by Eddington on the light of stars behind the sun during the eclipse of 1919.

[2] See Ehlers, J., Penrose, R. and Rindler, W. (1965) *Am. J. Phys.* **33**, 995, or the second edition (1966) of the author's *Special Relativity* (Oliver & Boyd).

8. Some four-momentum identities

We recall our definition (26.2) of four-momentum, three-momentum, and mass:

$$\mathbf{P} = m_0 \mathbf{U} = m_0 \gamma(u)(c, \mathbf{u}) = (mc, \mathbf{p}). \qquad (28.1)$$

From this we obtain the following equivalent expressions for $\mathbf{P}^2$:

$$\mathbf{P}^2 = m_0^2 c^2 = m^2 c^2 - p^2. \qquad (28.2)$$

Alternatively, the middle expression follows from the last by specializing to the rest frame.) And (28.2) can be rewritten, by use of $E = mc^2$, in the following two ways:

$$E^2 = p^2 c^2 + m_0^2 c^4, \quad p^2 = c^2(m^2 - m_0^2). \qquad (28.3)$$

When two particles with four-momenta $\mathbf{P}_1$ and $\mathbf{P}_2$ are involved, and their relative speed is v, we can establish the following identities:

$$\mathbf{P}_1 \cdot \mathbf{P}_2 = c^2 m_{01} m_2 = c^2 m_1 m_{02} = c^2 m_{01} m_{02} \gamma(v), \qquad (28.4)$$

where, typically, m_{01} is the rest mass of the first particle and m_2 is the mass of the second in the rest frame of the first. To obtain these, we evaluate $\mathbf{P}_1 \cdot \mathbf{P}_2$ in the rest frame of either particle. [As reference to (33.3) below shows, (28.4)(i) remains valid even if the second 'particle' is a photon; when both are photons, (28.4) becomes indeterminate and (33.2) below takes its place.]

Lastly consider an *elastic* collision between two particles, i.e. a collision in which the individual rest masses are preserved. Writing $\mathbf{P}$, $\mathbf{Q}$ for the pre-collision momenta and $\mathbf{P}'$, $\mathbf{Q}'$ for the post-collision momenta, we have $\mathbf{P} + \mathbf{Q} = \mathbf{P}' + \mathbf{Q}'$, which, upon squaring, gives $\mathbf{P}^2 + \mathbf{Q}^2 + 2\mathbf{P} \cdot \mathbf{Q} = \mathbf{P}'^2 + \mathbf{Q}'^2 + 2\mathbf{P}' \cdot \mathbf{Q}'$. But, by hypothesis, $\mathbf{P}^2 = \mathbf{P}'^2$ and $\mathbf{Q}^2 = \mathbf{Q}'^2$. It therefore follows that

$$\mathbf{P} \cdot \mathbf{Q} = \mathbf{P}' \cdot \mathbf{Q}'. \qquad (28.5)$$

This useful result we shall call the *elastic collision lemma*. (It remains valid even if one or both of the 'particles' are photons: a rebounding photon cannot help but preserve its rest mass which is always zero.) By (28.4)(iii), equation (28.5) is seen to be equivalent to the statement that the relative speed of the particles is the same before and after the collision. In this guise the lemma is also true in Newton's theory. That this is actually so can be directly deduced from the relativistic result $v = v'$ by letting $c \to \infty$. Why? Because in this limit the relativistics laws (26.5) and (27.1) as well as relativistic kinematics which govern collisions, go over into the Newtonian laws.]

29. Relativistic billiards

As a first example on the new mechanics we shall consider the relativistic analogue of a billiard ball collision—namely an elastic collision of two particles of equal rest mass, one of which is originally at rest. This analysis has many applications. Agreement with it provided the first direct confirmation of the relativistic collision law when Champion in 1932 bombarded stationary electrons in a cloud chamber with fast electrons from a radioactive source. A much more recent bubble-chamber experiment on elastic proton–antiproton scattering fits into the same framework.

Now if we approach this problem naively, setting up the conservation equations in the lab frame, we quickly get into a bad tangle of different γ-factors. So we look for a 'trick'. One would be to find an elegant four-vector argument, but here none presents itself naturally in spite of the availability of equation (28.5). The method we shall use instead is also of very general utility: we go to a frame where the problem is simpler, or even trivial, and then transform back to the frame of interest [cf. Exercise III(16)]. The frame where everything is obvious here is the so-called centre-of-momentum frame in which the two particles originally approach each other with equal and opposite constant velocities,[1] say $\pm v$ along the x'-axis. Call this frame S'. The only way to satisfy momentum and energy conservation in S' is for the post-collision velocities also to be $\pm v$, but possibly along some other line, say one making an angle θ' with the x'-axis (see Fig. 15). Now let S be in standard configuration with S' at velocity v. The 'right' particle is

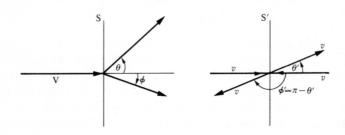

Fig.   15

then originally at rest in S, and so S is the required 'lab' frame. All we need to do now is to transform the remaining three velocities from S' to S. But in fact we are interested only in the *directions* of the post-collision velocities. So we can make use of the 'particle aberration' formula of Exercise III(13)—here needed in its inverse form $\tan \alpha = \sin \alpha'/\gamma(v)(\cos \alpha' + v/u')$. For the post-collision angles θ and ϕ in S, corresponding to θ' and $\phi' = \pi - \theta'$ in S', we then have

$$\tan \theta = \frac{\sin \theta'}{\gamma(v)(\cos \theta' + 1)}, \quad \tan \phi = \frac{\sin \theta'}{\gamma(v)(-\cos \theta' + 1)},$$

where we measure θ anticlockwise and ϕ clockwise. Multiplying these expressions together gives the first of the following equations

$$\tan \theta \, \tan \phi = \frac{1}{\gamma^2(v)} = \frac{2}{\gamma(V) + 1}. \qquad (29.1)$$

The second results when we apply (13.6)(ii) to the 'bullet', setting $u' = u_1' = v$ and $u = V$, V being the incident velocity in the lab. It is of course clear from momentum conservation in S that the angles θ and ϕ are coplanar.

Equation (29.1) gives the required relation between the incident velocity and the opening angle. The corresponding Newtonian result (here obtainable by letting $c \to \infty$, cf. end of Section 28) is $\tan \theta \tan \phi = 1$. This, as every billiard player knows, means $\theta + \phi = 90°$. $\left[\text{For it implies } \tan \theta = \cot \phi = \tan \left(\frac{\pi}{2} - \phi \right). \right]$ The relativistic formula implies that θ and ϕ are smaller than in the Newtonian case, i.e. that $\theta + \phi$ is less than 90°, and—on the average— progressively less as V increases (but for very small θ, ϕ could still be quite large).

---

[1] The possible existence of a short-range electric force between the particles does not affect the argument: details of what happens in the 'collision zone' are irrelevant.

30. The centre of momentum frame

We have seen in the preceding section how useful a frame can be in which the total three-momentum of a system of particles vanishes. Now we shall discuss the general case. Consider a frame S and in it a finite system of particles subject to no forces other than mutual collisions. We define its total mass $\bar{m}$, total momentum $\bar{\mathbf{p}}$, and total four-momentum $\overline{\mathbf{P}}$ as the *instantaneous* sums of the respective quantities of the individual particles:

$$\bar{m} = \sum m, \quad \bar{\mathbf{p}} = \sum \mathbf{p}, \quad \overline{\mathbf{P}} = \sum \mathbf{P} = \sum (mc, \mathbf{p}) = (\bar{m}c, \bar{\mathbf{p}}) \quad (30.1)$$

[cf. (26.2)]. Because of the conservation laws, each of the barred quantities remains constant in time.

The quantity $\overline{\mathbf{P}}$, being a sum of four-vectors, seems assured of four-vector status itself. But, in fact, it is not quite as simple as that. If all observers agreed on which $\mathbf{P}$s make up the sum $\Sigma \mathbf{P}$, then $\Sigma \mathbf{P}$ would clearly be a vector. But each observer forms this sum at one instant in *his* frame, which may result in different $\mathbf{P}$s making up the $\Sigma \mathbf{P}$ of different observers. A spacetime diagram such as Fig. 13 is useful in proving that $\Sigma \mathbf{P}$ is nevertheless a vector. (The suppression of the z-dimension makes the argument a little more transparent without affecting its general validity.) A simultaneity in S corresponds to a 'horizontal' plane π in the diagram and a simultaneity in a second frame S′ corresponds to a 'tilted' plane π'. In S, $\Sigma \mathbf{P}$ is summed over planes like π, and in S′ over planes like π'. However, we now assert that in S′ the same $\Sigma \mathbf{P}$ results whether summed over π' or π. For imagine a continuous motion of π' into π. As π' is tilted, each individual $\mathbf{P}$, located at a particle on π', remains constant (since the particles move uniformly between collisions) except when π' sweeps over a collision; but then the sub-sum of $\Sigma \mathbf{P}$ which enters the collision remains constant, by four-momentum conservation. Thus, without affecting the value, each observer *could* sum his $\mathbf{P}$s over the same plane π, and thus $\overline{\mathbf{P}}$ is indeed a four-vector.

Since $(\mathbf{P}_1 + \mathbf{P}_2 + \ldots)^2 = \mathbf{P}_1^2 + \ldots + 2\mathbf{P}_1 \cdot \mathbf{P}_2 + \ldots$, we see by reference to (28.2)(i) and (28.4), that the sum $\overline{\mathbf{P}}$ of any number of particle four-momenta is timelike, and it is also obviously future-pointing (first component positive). But then, by the result of the second paragraph of Section 22(v), we can always find a (clearly unique) frame S_{CM} moving relative to S with velocity

$$\mathbf{u}_{CM} = \frac{\bar{\mathbf{p}}}{\bar{m}}, \quad (30.2)$$

in which $\overline{\mathbf{P}}$ has no spatial cmponents, i.e. in which $\bar{\mathbf{p}} = 0$. For this reason S_{CM} is called the *centre of momentum* frame, or CM frame. We shall see later [cf. after (34.8)] that it is also the frame in which the centre of mass is at rest, though this needs some discussion. In relativity, in contrast to Newtonian mechanics, the centre of mass of a system is not uniquely determined. An example will show us why: suppose two identical particles move with opposite velocities along parallel lines l_1 and l_2 in some frame S, and suppose l is the line midway between these. In S the centre of mass lies on l. But in the rest frame of either particle the other particle is the more massive, and thus the centre of mass lies *beyond* l. So the centre of mass is frame-dependent. It is sometimes useful to define the *proper* centre of mass of a system as that in its CM frame.

Let us now write $\mathbf{U}_{CM}$ for the four-velocity $\gamma(u_{CM})(c, \mathbf{u}_{CM})$ of S_{CM}. Then, from (30.1) and (30.2),

$$\overline{\mathbf{P}} = (\bar{m}c, \bar{\mathbf{p}}) = \bar{m}(c, \mathbf{u}_{CM}) = \bar{m}\gamma^{-1}(u_{CM})\mathbf{U}_{CM}. \qquad (30.3)$$

On taking the norm of this four-vector equation, and writing $\bar{m}_{CM}$ for $\bar{m}$ in S_{CM}, we find

$$\sqrt{(\bar{m}^2 - \bar{p}^2/c^2)} = \bar{m}/\gamma(u_{CM}) = \bar{m}_{CM}, \qquad (30.4)$$

where the last equation results from specializing to the CM frame. Thus finally

$$\overline{\mathbf{P}} = \bar{m}_{CM}\mathbf{U}_{CM}. \qquad (30.5)$$

This equation shows that $\bar{m}_{CM}$ and $\mathbf{U}_{CM}$ are for the system what m_0 and $\mathbf{U}$ are for a single particle. They are the quantities that would be recognized as the rest mass and four-velocity of the system if its composite nature were ignored (as in the case of a macroscopic particle, which is made up of possibly moving molecules). Since the kinetic energy in S_{CM} is given by $T_{CM} = c^2(\bar{m}_{CM} - \bar{m}_0)$, where $\bar{m}_0 = \Sigma m_0$, we see that the effective rest mass of the system exceeds the rest masses of its parts by the mass equivalent of the kinetic energy of the parts in the CM frame. This is precisely what one would expect.

31. Threshold energies

An important application of relativistic mechanics occurs in so-called threshold problems. Consider, for example, the case of a free stationary proton (of rest mass M) being struck by a moving proton, whereupon not only the two protons but also a pion (of rest mass m) emerge. (Such reactions are often written in the form $p + p \rightarrow p + p + \pi^0$.) We shall ignore the electric interaction between the protons which is confined to a small collision zone. The question is, what is the minimum ('threshold') energy of the incident proton for this reaction to be possible? It is *not* simply $Mc^2 + mc^2$, i.e. it is *not* enough for the proton's kinetic energy to equal the rest energy of the newly created particle. For, by momentum conservation, the post-collision particles cannot be at rest, and so a part of the incident kinetic energy must *remain* kinetic energy. (It is a little like trying to smash ping-pong balls floating in space with a hammer.)

In all such cases the theoretical minimum expenditure of energy occurs when the end-products are *mutually* at rest. For consider, quite generally, two colliding particles: a 'bullet' and a stationary 'target', with respective pre-collision four-momenta $\mathbf{P}_1$ and $\mathbf{P}_2$. Let $\overline{\mathbf{P}}$ be the four-momentum of the post-collision system. Then $\mathbf{P}_1 + \mathbf{P}_2 = \overline{\mathbf{P}}$, whence

$$\mathbf{P}_1^2 + \mathbf{P}_2^2 + 2\mathbf{P}_1 \cdot \mathbf{P}_2 = \overline{\mathbf{P}}^2.$$

Substituting into this equation from (28.2), (28.4), and (30.5), we have

$$m_{01}^2 + m_{02}^2 + 2m_{01}m_{02}\gamma(v) = \bar{m}_{CM}^2. \tag{31.1}$$

Now the rest masses of bullet and target, m_{01} and m_{02}, are given. So the γ-factor of the bullet will be least when $\bar{m}_{CM}$ is least. That can never be *less* than $\bar{m}_{0(PC)}$, the sum of the *post-collision* (PC) rest masses; but it can *equal* $\bar{m}_{0(PC)}$ if the post-collision system travels as one lump. Hence the 'threshold' γ-factor (and so the threshold energy) will be given by (31.1) with $\bar{m}_{CM}$ replaced by $\bar{m}_{0(PC)}$:

$$m_{01}^2 + m_{02}^2 + 2m_{01}m_{02}\gamma(v) = \bar{m}_{0(PC)}^2. \tag{31.2}$$

[We shall presently see reasons why this formula, with $m_{01} = 0$ and $m_{01}\gamma(v) = h\nu/c^2$, applies even when the bullet is a photon; since it cannot travel as a lump with the other post-collision particles, it must be absorbed at threshold.]

For our original example of the proton–proton collision, equation (31.2) reads

$$M^2 + M^2 + 2M^2\gamma(v) = (2M + m)^2, \tag{31.3}$$

from which γ is immediate:

$$\gamma = 1 + \frac{2m}{M} + \frac{m^2}{2M^2}. \tag{31.4}$$

It is of some interest to look more closely at the efficiency of this and all analogous processes. The kinetic energy that must be supplied to the bullet (e.g. in a proton accelerator) is $(\gamma - 1)Mc^2$. The energy that actually goes into creating the new particle is mc^2. The ratio, k, of these energies is a measure of the efficiency. From the definition, and then from (31.3), we have

$$k = \frac{m}{(\gamma - 1)M} = \frac{2}{4 + (m/M)}. \tag{31.5}$$

Thus if the mass of the particle to be created is relatively small, the efficiency is close to 50 per cent. In our specific example, the ratio of the rest masses of pion to proton is ~ 0.14 and so $k \approx 48$ per cent. If we wanted to produce a proton–antiproton pair out of the collision of a proton and a neutron ($p + n \rightarrow p + n + p + \bar{p}$), the 'm' in our above analysis would be $2M$ (taking the masses of proton, antiproton and neutron as equal); so not only would this reaction require much more available energy than in the previous case (by a factor $2M/m$), but also the efficiency has now gone down to $1/3$. This is still not bad compared to what it *can* be in the case of relatively massive new particles. For example, the 'psi' particle (for whose discovery Richter and Ting received the Nobel prize in 1976) was created by the collision of electrons with positrons (antielectrons), whose rest mass it exceeds by a factor of 3700. The corresponding k would therefore be only $\sim 1/1850$,[1] requiring electron 'bullets' of energy $\sim 14\,000$ GeV, which exceeds the capacities of even the largest available electron accelerators by a factor of ~ 1000. So how was it done? It was done by a method that is almost 100 per cent efficient: the method of *clashing* (or *colliding*) *beams*. Here both target and bullet particles are first accelerated to high energy (for example, electrons and positrons can be accelerated in the same synchrotron, in opposite senses), then accumulated—sometimes for hours—in magnetic 'storage rings', before being finally loosed at each other head-on. No energy need be wasted, since the lab frame and the CM frame now coincide: *all* the kinetic energy can be used to create new matter.

[1] Since the electron–positron pair *annihilates*, (31.5) is now only approximately valid. The exact formula, $k = 2/(m/M - 4M/m)$, results from omitting $2M$ on the right-hand side of (31.3).

32. De Broglie waves

In Section 24 we discussed the dualistic relation between photons (rays) and light-waves and, more generally, that between arbitrary waves and their rays. In this section we shall add a dynamic dimension to our earlier kinematic discussion, and uncover the general particle–wave dualism which lies at the root of quantum mechanics, but whose very possibility depends on *relativistic* kinematics and dynamics.

Guided by thermodynamic considerations, Planck in 1900 had made the momentous suggestion that radiant energy is *emitted* in definite 'quanta' of energy

$$E = hv, \tag{32.1}$$

where v is the frequency of the radiation and h a universal constant (Planck's constant). In 1905 Einstein, crystallizing the observed facts of the photoelectric effect, suggested that not only is radiant energy *emitted* thus, but that it also travels and is absorbed as quanta, which were later called photons. According to Einstein, a photon of frequency v has energy hv, and thus a finite mass hv/c^2 and a finite momentum hv/c. It can be regarded as a limiting particle travelling at speed c, for which the mass γm_0 is finite, while $\gamma = \infty$ and $m_0 = 0$.

In 1923 de Broglie proposed a further extension of this idea, namely that associated with *any* particle of energy E there are waves of frequency E/h travelling in the same direction. However, as we have seen at the end of Section 24, these waves cannot travel at the same speed as the particle (unless that speed is c), for then the association would not be Lorentz-invariant. De Broglie found that, for consistency, the speeds u and w of the particle and its associated wave, respectively, must be related by the equation

$$uw = c^2. \tag{32.2}$$

Let us see why. De Broglie's proposal was to associate with a particle of four-momentum **P** a wave whose frequency four-vector **N** [cf. (24.3)] is related to **P** by the following equation, now known as *de Broglie's equation:*

$$c\mathbf{P} = h\mathbf{N}, \quad \text{i.e. } cm(c, \mathbf{u}) = hv(1, c\mathbf{n}/w). \tag{32.3}$$

Being a four-vector equation, it is certainly Lorentz-invariant. But moreover, if Plancks relation $hv = E(= mc^2)$ *does* generalize to an arbitrary particle and its associated wave, then de Broglie's equation is in fact inevitable. For the vector $c\mathbf{P} - h\mathbf{N}$ will then have its first component zero in all inertial frames, and so, by the 'zero component' lemma of Exercise IV(5), it must vanish entirely. And then, equating components in (32.3), we find

$$m = \frac{hv}{c^2}, \quad \mathbf{u} \propto \mathbf{n}, \quad mu = \frac{hv}{w}. \tag{32.4}$$

The first of these equations is Planck's relation; the second locks the wave normal to the direction of motion of the particle; but the third (with the first) shows the inevitability of (32.2).

Note fom (32.1) that v can neither vanish nor be infinite, even if the particle is at rest (when $u = 0$ and $w = \infty$). But the wavelength, $\lambda = w/v$, is infinite for a particle at rest.

For particles of non-zero rest mass, $u < c$ and thus $w > c$. The de Broglie wave speed then has an interesting and simple interpretation. Suppose a whole swarm of identical particles travel with equal velocity, and something happens to all of them simultaneously in their rest frame: suppose, for example, they all 'flash'. Then this flash sweeps over the particles at the de Broglie velocity in any other frame. To see this, suppose the particles are at rest in the usual frame S', travelling at speed v relative to a frame S. Setting $t' = 0$ for the flash, we find from (6.14)(i) that $x/t = c^2/v$, and this is evidently the speed of the flash in S. Thus de Broglie waves can be regarded as 'waves of simultaneity'.

De Broglie's idea proved to be of fundamental importance in the development of quantum mechanics. It is one of the profound discoveries brought to light by special relativity, almost on a par with $E = mc^2$. One of its first successes was in explaining the permissible electron orbits in the old Bohr model of the atom, as those containing a whole number of electron waves. A striking empirical verification of electron waves came in 1927, when Davisson and Germer first observed the phenomenon of electron diffraction. And, of course, the superiority of the electron microscope hinges on de Broglie's relation, according to which electrons allow us to 'see' with very much greater resolving power than photons since they have very much smaller wavelengths. Another satisfactory feature of de Broglie's hypothesis is that it explains certain analogies between mechanics and optics that had previously appeared to be purely coincidental—for example, that between Maupertuis' principle of least action and Fermat's principle of least time, or that between Jacobi's and Hamilton's techniques and ray optics. In fact, classical mechanics can now be regarded as an approximation to a more accurate *wave mechanics* in much the same way that geometrical optics (ray optics) is an approximation to physical optics (wave optics). For classical problems the approximation is very good since the wavelengths associated with macroscopic particles is minute. But when the dimensions of the problem become comparable to the wavelengths, then, as in optics, the approximation can be expected to fail.

33. Photons

In the special case of light, (32.2) is satisfied by $u = w = c$, and from (32.3) we then find for the four-momentum of a photon

$$\mathbf{P} = \frac{h\nu}{c}(1, \mathbf{n}).\tag{33.1}$$

But this can also be wtitten down directly from the definition (26.2)(iii) and Einstein's values $h\nu/c^2$ and $h\nu/c$ for the mass and momentum of a photon. Clearly (33.1) is a *null* vector: $\mathbf{P}^2 = 0$.

We shall assume that, in collision problems, photons can be treated like any other particles, and that conservation of energy and momentum also apply to particle systems with photons. We recall (cf. Section 30) that for the existence of a 'CM frame' the total four-momentum $\overline{\mathbf{P}}$ of a system must be timelike and future-pointing. Accordingly, systems consisting of photons only may or may not have a CM frame (a single photon certainly does *not*); but any system containing at least one particle of non-zero rest mass, or two non-parallel photons, *does* have a CM frame. (The reader may recall a relevant heuristic argument at the end of the penultimate paragraph of Section 21.) To establish this result formally, consider first two photons with four-momenta $\mathbf{P}_1$ and $\mathbf{P}_2$, frequencies ν_1 and ν_2, and velocities that make an angle θ with each other. Then, from (33.1),

$$\mathbf{P}_1 \cdot \mathbf{P}_2 = h^2 c^{-2} \nu_1 \nu_2 (1 - \cos\theta),\tag{33.2}$$

which is positive unless $\theta = 0$, i.e. unless the photons are parallel. Next consider a photon and a particle of non-zero rest mass m_0, with respective four-momenta $\mathbf{P}$ and $\mathbf{Q}$. Then, evaluating $\mathbf{P} \cdot \mathbf{Q}$ in the rest frame of the particle, we find

$$\mathbf{P} \cdot \mathbf{Q} = h\nu_0 m_0,\tag{33.3}$$

where ν_0 is the frequency of the photon in that frame. The product is clearly positive. By use of (33.2) and (33.3) we can now generalize our previous argument leading to (30.2) and so to the existence of a (unique) CM frame. Thus our result is established.

Equations (33.2) and (33.3) have relevance also to our earlier formulae (28.4) and (31.2): they now provide the justification for the remarks we made after those formulae about their applicability to photons.

An experiment that was instrumental in getting the corpuscular theory of light widely accepted was Compton's famous X-ray scattering experiment of 1922. It was the first to show photons behaving 'like billiard balls', being bounced off other particles (electrons) in full conformity with the relativistic laws of collision mechanics. And thereby it incidentally was also the first to test those laws. Let us now apply our formalism to its analysis. We can use the left half of Fig. 15 to illustrate the situation: a photon of frequency v is incident along the x-axis, strikes a stationary electron, and scatters at an angle θ with diminished frequency v' while the electron recoils at an angle ϕ. We wish to relate v, v' and θ with m, the rest mass of the electron. (No observations were made, at first, on the recoiling electron.) If $\mathbf{P}$, $\mathbf{P}'$ are the pre- and post-collision four-momenta of the photon and $\mathbf{Q}$, $\mathbf{Q}'$ those of the electron, then $\mathbf{P} + \mathbf{Q} = \mathbf{P}' + \mathbf{Q}'$, by conservation of four-momentum. Now, following a pleasant method of Lightman et al.,[1] we isolate the unwanted vector $\mathbf{Q}'$ on one side of the equation and square to get rid of it:

$$(\mathbf{P} + \mathbf{Q} - \mathbf{P}')^2 = \mathbf{Q}'^2.$$

Since $\mathbf{Q}^2 = \mathbf{Q}'^2$ and $\mathbf{P}^2 = \mathbf{P}'^2 = 0$, we are left with $\mathbf{P} \cdot \mathbf{Q} - \mathbf{P}' \cdot \mathbf{Q} - \mathbf{P} \cdot \mathbf{P}' = 0$, i.e.

$$\mathbf{P} \cdot \mathbf{P}' = \mathbf{Q} \cdot (\mathbf{P} - \mathbf{P}'), \tag{33.4}$$

from which we find at once, by reference to (33.2) and (33.3), the desired relation

$$hc^{-2}vv'(1 - \cos\theta) = m(v - v'). \tag{33.5}$$

In terms of the corresponding wavelengths λ, λ', and the half-angle $\theta/2$, this may be rewritten in the more standard form

$$\lambda' - \lambda = (2h/cm)\sin^2(\theta/2). \tag{33.6}$$

Scattering of photons by stationary electrons is called *Compton scattering*, and clearly always results in a loss of energy to the photon. The opposite is true for *inverse Compton scattering*, in which a photon collides with a fast ('relativistic') electron or other charged particle and often experiences a spectacular gain in energy.[2] This process is an important source of intergalatic X-rays, but it also has applications in the laboratory.

Compton scattering and inverse Compton scattering are, of course, identical processes in their CM frames. Thus one way to obtain the result of an inverse Compton collision is to transform a Compton collision to a rapidly moving frame, using the transformation properties of photons [cf. Exercise V(19)]. But it is just as easy to proceed *ab initio*, by specializing equation (33.4) to the present frame of interest (e.g. intergalactic space), in which the incident electron moves relativistically, i.e. such that $\gamma \gg 1$. Relative to that frame, let us put

$$\mathbf{Q} = \gamma m(1, \mathbf{u}), \quad \mathbf{P} = h\nu(1, \mathbf{n}), \quad \mathbf{P}' = h\nu'(1, \mathbf{n}'),$$

(*now using units such that* $c = 1$), and let us consider only the extreme case of a head-on collision, so that

$$\mathbf{u} \cdot \mathbf{n} = -u, \quad \mathbf{u} \cdot \mathbf{n}' = \mathbf{u}, \quad \mathbf{n} \cdot \mathbf{n}' = -1.$$

Then (33.4) reads

$$2h^2 \nu\nu' = \gamma m h\nu(1 + u) - \gamma m h\nu'(1 - u), \tag{33.7}$$

from which we find, solving for $h\nu'$,

$$h\nu' = \frac{(\gamma + \gamma u)m}{2 + (\gamma - \gamma u)m/h\nu} \approx \frac{\gamma m}{1 + m^2/4(\gamma m)(h\nu)}, \tag{33.8}$$

where the second expression results from the approximation $\gamma u = (\gamma^2 - 1)^{1/2} \approx \gamma - 1/2\gamma$. (It is, of course, *a priori* evident that for 'large' u, i.e. $u \approx 1$, $\gamma u \approx \gamma$.)

As an example, consider a photon of the 3 K cosmic 'background' radiation being hit by a high-energy proton, say one with $\gamma = 10^{11}$.[3] The photon's energy $h\nu$ is of the order of Boltzmann's constant times 3 K, which is $\sim 3 \times 10^{-4}$ eV.[4] With $m \approx 10^9$ eV, our formula then yields $\sim 10^{19}$ eV (!) for the energy $h\nu'$ of the scattered photon. Clearly our proton cannot suffer many such collisions without being slowed considerably. If its energy were ten times higher, half of it would be lost in a single collision.

It may be noted that when, as in our example, the product $(\gamma m)(h\nu)$ of the pre-collision energies is small compared to the square of the particle's rest energy, m^2, (33.8) can be written in the useful form

$$\frac{h\nu'}{h\nu} \approx 4\gamma^2. \tag{33.9}$$

Two other related mechanisms for creating radiation—both in the cosmos and in the laboratory—are *bremsstrahlung* (German for breaking radiation), in which an electron collides elastically with a nucleus and emits a photon, and *magnetic bremsstrahlung* (or *synchrotron radiation*), in which the electron is deflected not by a collision but by a magnetic field. From the point of view of the electron (or other charged particle), Compton scattering, bremsstrahlung, and synchrotron radiation are just different mechanisms for producing the same effect—a velocity change of the electron, which *therefore* radiates (in obedience to Maxwell's equations, as we shall see in Chapter VI). That the first two can be treated correctly by collision mechanics is really a remarkable fact, and suggests that the range of validity of energy and momentum conservation extends to electromagnetic fields.

[1] Lightman, A. P., Press, W. H., Price, R. H., and Teukolsky, S. A. (1975) *Problem Book in Relativity and Gravitation*, Princeton University Press, p. 159.

[2] That the bullet must be *charged* for such a 'collision' to occur is a consequence of the electromagnetic nature of the radiation (light) represented by the photon.

[3] Taken from Lightman *et al.* (1975). See note [1] above.

[4] Energies of elementary particles are usually measured in electron-volts, eV. One eV is the kinetic energy gained by an electron (or proton) in traversing a potential difference of one volt. In fact, $1 \, eV = 1.602 \times 10^{-19}$ joule. Bigger units are the MeV ($= 10^6$ eV: 'M' for mega) and the GeV ($= 10^9$ eV: 'G' for *giga* = big). Rest energies of particles are also expressed in eV. For example, that of the electron is about 0.511 MeV and that of the proton 0.938 GeV.

34. The angular momentum four-tensor

The present section continues the general discussion begun in Section 30 on the collision mechanics of otherwise free systems of particles. Recall the Newtonian definition of the angular momentum

$$\mathbf{h} = \mathbf{r} \times \mathbf{p} \qquad (34.1)$$

of a single particle having position vector $\mathbf{r}$ relative to some given origin, and linear momentum $\mathbf{p}$. We shall adopt the same definition in relativity, except that $\mathbf{p}$ shall from now on mean the *relativistic* momentum. If $x^\mu = (ct, \mathbf{r})$ and $P^\mu = m_0 U^\mu = (mc, \mathbf{p})$ are the corresponding four-vectors (under homogeneous Lorentz transfor-

mations), we define the (antisymmetric) *angular momentum four-tensor*
$L^{\mu\nu}$ of that particle, relative to the spacetime origin, by the equation

$$L^{\mu\nu} = x^{\mu}P^{\nu} - x^{\nu}P^{\mu}. \tag{34.2}$$

That it *is* a four-tensor is evident by its mode of formation from the
two four-vectors x^{μ} and P^{μ}. For its space–space part we have
$(L^{23}, L^{31}, L^{12}) = (h_1, h_2, h_3)$, and so

$$L^{ij} = e^{ijk}h_k, \quad h_i = \tfrac{1}{2}e_{ijk}L^{jk}, \tag{34.3}$$

where e_{ijk} and e^{ijk} are the permutation symbols defined in Exercises
A(15) and A(16).

Now consider a system of particles subject to no forces other than
mutual collisions, as in Section 30. We define its *total* angular
momentum four-tensor $\overline{L}^{\mu\nu}$ as the instantaneous sum of the $L^{\mu\nu}$ of the
separate particles:

$$\overline{L}^{\mu\nu} = \sum L^{\mu\nu}. \tag{34.4}$$

But just as in the case of the total four-momentum $\overline{P}^{\mu}$, the question
arises whether $\overline{L}^{\mu\nu}$ is indeed a four-tensor. Of course all the summands
in (34.4) are tensors, but the summation in different frames extends
over different simultaneities. However, just as in the case of $\overline{P}^{\mu}$, we can
show that it makes no difference over *which* simultaneities $\overline{L}^{\mu\nu}$ is
summed.

To this end consider first a single particle, with proper time τ and
four-velocity U^{μ}, in between collisions. Then $P^{\mu} =$ constant, whence
from (34.2),

$$(d/d\tau)L^{\mu\nu} = m_0(U^{\mu}U^{\nu} - U^{\nu}U^{\mu}) = 0,$$

which shows that $L^{\mu\nu}$ is constant along the particle's worldline. Next
consider a group of particles *during* a collision in which they all
participate. Since they are all at the same place at the instant of the
collision, the sum of their angular momenta just before and just after
collision takes the special form

$$\sum L^{\mu\nu} = x^{\mu}\sum P^{\nu} - x^{\nu}\sum P^{\mu}.$$

But $\sum P^{\mu}$ is unchanged by the collision, and so at each collision the
relevant sub-sum of $\overline{L}^{\mu\nu}$ is conserved. We can now repeat, *mutatis
mutandis*, the 'plane tilting' argument that established the tensor
character of $\overline{P}^{\mu}$ in Section 30. Thus $\overline{L}^{\mu\nu}$ *is a four-tensor, and it is
constant in time*.

The significance of the space–space part of $\overline{L}^{\mu\nu}$ is evident from (34.3) and (34.4):

$$\overline{L}^{ij} = e^{ijk}\overline{h}_k, \tag{34.5}$$

where $\overline{h}_k = \Sigma h_k$ is the *total three-angular momentum of the system, which is therefore conserved.* For the time–space part at any instant t we have, from (34.2) and (34.4),

$$\overline{L}^{0i} = ct \sum p^i - c \sum mx^i = c(t\,\overline{p}^i - \overline{m}x_C^i), \tag{34.6}$$

where $\overline{p}^i = \Sigma p^i$, $\overline{m} = \Sigma m$, and

$$x_C^i = \sum mx^i/\overline{m}. \tag{34.7}$$

The latter is the position vector of the *centroid*—the centre of mass— in the inertial frame under consideration. Since $\overline{L}^{0i}$ and $\overline{p}^i$ are constant, we find, by differentiating (34.6),

$$\frac{dx_C^i}{dt} = \frac{\overline{p}^i}{\overline{m}}, \tag{34.8}$$

which shows that *the centroid moves uniformly.* Moreover, by comparing (34.6) with (30.2), we see that it moves with the same velocity as S_{CM}, and is therefore at rest in S_{CM}, which bears out the assertion we made after (30.2). [A more direct proof is suggested in Exercise V(8).] Note the remarkable linkage of the 'centroid theorem' with three-angular momentum conservation in the relativistic law of four-angular momentum conservation. In fact, we cannot have the one without the other—by an obvious extension of the 'zero component' lemma [see Exercise VI(7) below].

Next consider the effect on $\overline{L}^{\mu\nu}$ of a change of 'pivot' $O \to y^\mu$. Then, in an obvious notation,

$$\overline{L}^{\mu\nu}(y) = \sum \left[(x^\mu - y^\mu)P^\nu - (x^\nu - y^\nu)P^\mu \right]$$
$$= \overline{L}^{\mu\nu}(O) - (y^\mu \overline{P}^\nu - y^\nu \overline{P}^\mu). \tag{34.9}$$

In the CM frame (where $\overline{p}^i = 0$) we have, in particular,

$$\overline{L}_{CM}^{ij}(y) = \overline{L}_{CM}^{ij}(O). \tag{34.10}$$

Since $\overline{h}_i = \frac{1}{2}e_{ijk}\overline{L}^{jk}$ [cf. (34.3)], this means that *the three-angular momentum in the CM frame,* $\mathbf{h}_{CM}$, *is independent of the pivot.* We can therefore regard $\mathbf{h}_{CM}$ as the intrinsic or 'spin' angular momentum of the system.

An important four-vector in this connection is the (*Pauli–Lubanski*) *spin vector*, defined by

$$S_\mu = \frac{1}{2c}\varepsilon_{\mu\nu\rho\sigma} \overline{L}^{\nu\rho} V^\sigma, \tag{34.11}$$

where V^σ is the four-velocity of the CM frame (and is thus parallel to $\overline{P}^\sigma$). By use of (34.9) it can be verified that S_μ is independent of the pivot, and so it must represent the intrinsic angular momentum $\overline{\mathbf{h}}_{CM}$. In fact, in the CM frame $[V^\sigma = (c, \mathbf{0})]$ S^μ reduces to $(0, \overline{\mathbf{h}}_{CM})$.

If, in the general frame, we choose $x_C^\mu = (ct, x_C^i)$ for y^μ, we find from (34.9) and (34.6) that

$$\overline{L}^{0i}(x_C) = 0. \tag{34.12}$$

(An equation like this could not possibly hold if the centroid were the same in all frames! For whereas a tensor equation like $T^{\mu\nu} = 0$ is form-invariant, a component equation like $T^{0i} = 0$ is not.) Equation (34.12) leads directly to the *Fokker–Synge equation*

$$\overline{P}_\mu \overline{L}^{\mu\nu}(x_{PC}) = 0, \tag{34.13}$$

where x_{PC} refers to the *proper centroid*, i.e. the centroid in the CM frame. The validity of the tensor equation (34.13) becomes apparent when we specialize it to the CM frame, where (34.12) holds for x_{PC} and $\overline{P}_i = 0$. The chief interest of (34.13) lies in the fact that, conversely, it *uniquely* determines the worldline of the proper centroid. For let us look at (34.13) in the CM frame, where it reduces to (34.12) with $x_C = x_{PC}$, and choose spatial coordinates so that $x_{PC}^i = 0$. Now suppose there exists another pivot y for which $\overline{L}^{0i}(y) = 0$. But then (34.9) shows at once that $y^i = 0$ (since we must assume $\overline{P}^0 = \bar{m}c \neq 0$), and our assertion is proved.

Consider, for a moment, what would happen if a finite system were to shrink to a point. Now, whereas such a limiting procedure leads to no 'trouble' in Newtonian mechanics, in relativistic mechanics it does, and the 'trouble' resides in equation (34.12). For, if all mass is positive, every observer must place the centroid of a particle *at* the particle, which means that (34.12) would have to hold universally for each point x on the particle's worldline, and we have already seen reasons why this cannot be.[1] In addition, a system of free particles would first have to be endowed with cohesive forces in order to make the limit possible. Nevertheless, without regarding it as the limit of a finite system, one can usefully introduce the concept of a relativistic point-particle with spin (which is denoted with a different letter, $S^{\mu\nu}$, to indicate lack of complete analogy with $L^{\mu\nu}$). Such a particle is assumed to possess three things: (i) a straight worldline l representing the history of its centroid; (ii) a four-momentum P^μ parallel to l; and (iii) an antisymmetric spin angular momentum tensor $S^{\mu\nu}$ with respect to any point on l, satisfying $P_\mu S^{\mu\nu} = 0$, i.e. equation (34.13), which *can* be taken over from finite systems. It is then easy to see that if, in a system of such particles, the spin angular momentum of each particle is conserved *between* collisions, and the total spin angular momentum is conserved *in* each collision, the conservation of *total* angular momentum $\Sigma(L^{\mu\nu} + S^{\mu\nu})$ is assured, as well as its tensor character.

[1] This argument suggests that there must be a minimum to the volume that a particle system with given $\overline{P}^\mu$ and $\overline{L}^{\mu\nu}$ can occupy. This minimum turns out to have at least one radius as big as $|\overline{\mathbf{h}}_{CM}|/c\overline{m}_{CM}$. [For an excellent discussion of angular momentum in special relativity, and this point in particular, see W. G. Dixon, *loc. cit.* end of Section 4.] Of course, intuitively it is easy to see why: if the dimensions of the system are decreased, there must come a stage when the outer particles have to move at the speed of light [cf. Exercise V (24)].

35. Three-force and four-force

We have come quite a long way in our development of relativistic mechanics without the need to introduce the concept of force. But that, too, is important in many situations which call for relativistic mechanics, such as the motions of fast elementary particles in electromagnetic fields. Recall the two familiar forms of Newton's second law:

$$\mathbf{f} = m\mathbf{a}, \quad \mathbf{f} = d\mathbf{p}/dt. \tag{35.1}$$

Strictly speaking, this is only 'half' a law; for it is a mere definition of force, devoid of physical content, until it is combined with another such 'half-law', e.g. the law of gravity $\mathbf{f} = -Gm_1 m_2 r^{-3}\mathbf{r}$, or Lorentz's force law $\mathbf{f} = q(\mathbf{e} + \mathbf{u} \times \mathbf{b})$. Then the two half-laws together give the motion of the particle. So to some extent the definition of force is a matter of convenience, and must be guided by other half-laws involving force.

Our first 'guess' in Section 26 for the definition of four-force was

$$\mathbf{F} = m_0 \mathbf{A} = m_0 \frac{d\mathbf{U}}{d\tau}, \tag{35.2}$$

modelled on the first form of Newton's law. But it turns out that a definition modelled on the *second* form of that law is the one that connects naturally with other important 'half-laws' about force, and which has subsequently been borne out by experience:

$$\mathbf{F} = \frac{d\mathbf{P}}{d\tau} = \frac{d}{d\tau}(m_0 U). \tag{35.3}$$

Both (35.2) and (35.3) are manifestly tensors. But they are equivalent *only* if the rest mass is constant, which is by no means always the case. For example, if two particles collide elastically, their rest masses *during* collision will vary, but that would be precisely when we might be interested in the elastic forces acting on them. In fact, we shall see presently that (35.3) allows us to 'prove' Newton's third law of the equality of action and reaction for two particles in collision. (For particles interacting at a distance the question of simultaneity vitiates all attempts to incorporate Newton's third law into relativity.)

If we define the *(relativistic) three-force* by

$$\mathbf{f} = \frac{d\mathbf{p}}{dt} = \frac{d(m\mathbf{u})}{dt}, \tag{35.4}$$

where $\mathbf{p}$, m are, of course, the *relativistic* momentum and mass of the particle on which $\mathbf{f}$ acts, we can write (35.3) in the form [cf. (26.2) and (23.2)]

$$\mathbf{F} = \gamma(u)\frac{d}{dt}(mc, \mathbf{p}) = \gamma(u)\left(c\,\frac{dm}{dt}, \mathbf{f}\right). \tag{35.5}$$

No comparably simple relation would exist between an $\mathbf{F}$ as defined in (35.2) and any 'reasonably' defined $\mathbf{f}$. Once again nature seems to choose the formally simple path.

The appearance of $\gamma\mathbf{f}$ as the spatial part of a four-vector immediately tells us the transformation law of $\mathbf{f}$. By reference to (22.4) and (13.6) (i) we find, with $\gamma = \gamma(v)$,

$$f_1' = \frac{f_1 - v\,dm/dt}{1 - u_1 v/c^2} \qquad \left(f_1' = \frac{f_1 - v\mathbf{f}\cdot\mathbf{u}/c^2}{1 - u_1 v/c^2}, \text{ if } m_0 = \text{const}\right)$$

$$f_2' = \frac{f_2}{\gamma(1 - u_1 v/c^2)}, \quad f_3' = \frac{f_3}{\gamma(1 - u_1 v/c^2)}, \tag{35.6}$$

where the second formula in the first row applies when m_0 = constant, as we shall show presently. Note the formal similarity of the force transformation (35.6) with the velocity transformation (13.3), which is explained by the similarity of (35.5) with (23.4). Observe also that the transformation of the three-force depends on the velocity of the particle on which it acts. Unlike its Newtonian counterpart, the relativistic $\mathbf{f}$ is not invariant between frames. But when m_0 is constant and $\mathbf{f}$ is parallel to the particle's velocity $\mathbf{u}$, then $\mathbf{f}$ is invariant among all frames whose motion is also parallel to $\mathbf{u}$. This follows from the parenthesized formula on putting $u_1 = u$ and $f_1 = f$, $f_2 = f_3 = 0$; it follows even more readily by putting $u = 0$, i.e. by transforming from the particle's *rest* frame to the general frame.

We are now ready to prove Newton's third law in the case of contact. In some particular frame S let the four-forces exerted by two colliding particles on each other be $\mathbf{F}_1$ and $\mathbf{F}_2$, with

$$\mathbf{F}_a = \gamma(u_a)\left(c\,\frac{dm_a}{dt}, \mathbf{f}_a\right), \quad (a = 1, 2).$$

Then, since the particles are in contact, $\mathbf{u}_1 = \mathbf{u}_2$, and, by energy conservation, $d(m_1 + m_2)/dt = 0$. Consequently the first component of the vector $\mathbf{F}_1 + \mathbf{F}_2$ vanishes in S, and, by the same argument, in all other inertial frames. But then, by the 'zero-component' lemma of Exercise IV(5), it follows that $\mathbf{F}_1 + \mathbf{F}_2 = 0$. This, in turn, implies $\mathbf{f}_1 + \mathbf{f}_2 = 0$, which establishes Newton's third law for particles in contact. This must surely be regarded as support for the definitions (35.3) and (35.4). We should not be too surprised that in our development Newton's law appears as a theorem; after all, we took momentum conservation as an axiom, thus reversing Newton's logical sequence.

Now consider a particle with four-velocity $\mathbf{U} = \gamma(c, \mathbf{u})$ being subjected to a four-force $\mathbf{F} = \gamma(cdm/dt, \mathbf{f})$. Then

$$\mathbf{U} \cdot \mathbf{F} = \gamma^2\left(c^2\frac{dm}{dt} - \mathbf{f}\cdot\mathbf{u}\right) = c^2\frac{dm_0}{d\tau} = \gamma c^2\frac{dm_0}{dt}, \qquad (35.7)$$

where the second equation results from specializing the first to the rest frame of the particle, and the third is a consequence of (23.2). From (35.7) we get

$$c^2\frac{dm}{dt} = \mathbf{f}\cdot\mathbf{u} + \frac{c^2}{\gamma}\frac{dm_0}{dt}, \qquad (35.8)$$

which immediately justifies the parenthesized equation in (35.6) as a special case of that preceding it.

We shall call a force *pure* if it does not change a particle's rest mass, and *heatlike* if it does not change a particle's velocity. For observe from (35.4) that a particle which is at rest but being heated (e.g. by a candle held under it) in one frame, experiences a 'force' $\mathbf{f} = (dm_0/dt)\gamma(u)\mathbf{u}$ in every other frame. The necessary and sufficient condition for a force to be pure follows at once from (35.7):

$$\mathbf{U} \cdot \mathbf{F} = 0 \Leftrightarrow m_0 = \text{constant}, \tag{35.9}$$

while that for a force to be heatlike is, of course, $\mathbf{A} = 0$. It will turn out that one of the most important forces, the Lorentz force of Maxwell's theory, is pure. On the other hand, the action and reaction forces that occur during a collision are heatlike. An example of an *impure* force is provided by one that is derivable from a four-scalar potential Φ according to the equation

$$F_\mu = \partial \Phi / \partial x^\mu. \tag{35.10}$$

(For its tensor character cf. Section A6.) This equation implies

$$\mathbf{U} \cdot \mathbf{F} = \frac{\partial \Phi}{\partial x^\mu} \frac{dx^\mu}{d\tau} = \frac{d\Phi}{d\tau},$$

which vanishes only in the trivial case $\Phi = \text{constant}$. Otherwise, in conjunction with (35.7) (ii), it leads to $c^2 m_0 = \Phi + \text{constant}$. In fact, the scalar meson theory of the nucleus assumes that the constituent nucleons are held together by an attractive short-range force of just this type.

In the case of a *pure* force $\mathbf{f}$, we can define dW, the work done by the force in moving its 'point of application' through $d\mathbf{r}$, by the Newtonian equation

$$dW = \mathbf{f} \cdot d\mathbf{r}. \tag{35.11}$$

For by (35.8) this implies $dW = d(mc^2)$, i.e. work done equals increase in energy. Here we have another recommendation for the definition (35.4) of $\mathbf{f}$. But (35.11) does *not* apply when the force is impure, for then it gives neither the total energy increment $d(mc^2)$, nor the kinetic energy increment $d(m - m_0)c^2$, as is clear from (35.8).

Again, for a *pure* force, we have, from (35.3), (35.5), and (35.8),

$$\mathbf{F} = m_0 \mathbf{A} = \gamma(u)\left(\frac{1}{c}\mathbf{f} \cdot \mathbf{u}, \mathbf{f}\right), \tag{35.12}$$

and from (35.4) and (35.8),

$$\mathbf{f} = m\mathbf{a} + \frac{dm}{dt}\mathbf{u} = \gamma m_0 \mathbf{a} + \frac{\mathbf{f} \cdot \mathbf{u}}{c^2}\mathbf{u}. \qquad (35.13)$$

We shall discuss this last formula in a little detail. Although $\mathbf{u}$, $\mathbf{a}$, and $\mathbf{f}$ are coplanar, the acceleration is not in general parallel to the force. Evidently there are just two cases when it *is*, namely when $\mathbf{f}$ is either parallel or orthogonal to $\mathbf{u}$. In particular, in the particle's rest frame we have $\mathbf{f} = m_0\mathbf{a}$. It is instructive to consider the components of $\mathbf{f}$ and $\mathbf{a}$ along $\mathbf{u}$ and along a unit vector $\mathbf{n}$ orthogonal to $\mathbf{u}$ and coplanar with $\mathbf{u}$, $\mathbf{a}$, $\mathbf{f}$. Forming the scalar product of (35.13) with $\mathbf{u}/u$, we find

$$f_\parallel = \gamma m_0 a_\parallel + f_\parallel u^2/c^2, \quad \text{i.e.} \quad f_\parallel = \gamma^3 m_0 a_\parallel, \qquad (35.14)$$

where $f_\parallel = \mathbf{f} \cdot \mathbf{u}/u$ and $a_\parallel = \mathbf{a} \cdot \mathbf{u}/u$. Similarly, multiplying (35.13) with $\mathbf{n}$ gives

$$f_\perp = \gamma m_0 a_\perp, \qquad (35.15)$$

where $f_\perp = \mathbf{f} \cdot \mathbf{n}$ and $a_\perp = \mathbf{a} \cdot \mathbf{n}$. It appears, therefore, that a moving particle offers different inertial resistances to the same force according to whether it is subjected to that force longitudinally or transversely. In this way there arose the (now somewhat outdated but still often useful) concepts of 'longitudinal' mass $\gamma^3 m_0$, and 'transverse' mass γm_0. Since any force can be resolved into longitudinal and transverse components, equations (35.14) and (35.15) provide one method of calculating the resultant acceleration.

36. Relativistic analytic mechanics

We end this chapter with a brief discussion of the relativistic form of Hamilton's principle and related matters, which, however, can be omitted at a first reading. The subject is of interest, for example, in connection with doing relativistic problems in non-standard coordinates, and with relativistic quantum theories.

Suppose a particle of rest mass m_0 moves with four-velocity U^μ under the action of a four-force F^μ. Let us further suppose that F^μ is derivable from a four-*vector* potential Φ_μ according to the equation

$$F_\nu = \frac{q}{c}(\Phi_{\mu,\nu} - \Phi_{\nu,\mu})U^\mu, \qquad (36.1)$$

where q is a constant. [For the notation cf. Section A9.] Such a force is necessarily 'pure', i.e. rest mass preserving, for [cf. (35.9)]

$$F_\nu U^\nu = \frac{q}{c}(\Phi_{\mu,\nu} - \Phi_{\nu,\mu})U^\mu U^\nu = 0.$$

A particular case of (36.1) corresponds to motion under no force at all ($\Phi_\mu \equiv 0$). Another corresponds to motion under a force F^μ whose associated three-force is derivable from a scalar potential $\phi(ct, x, y, z)$ in one specific frame (this is *not* a Lorentz invariant relation). For if we set $\Phi_\mu = (\phi, 0, 0, 0)$ in that frame, then (36.1) gives the first of the following equations

$$F_\mu = \frac{q}{c}\gamma(-\phi_{,i}u^i, \phi_{,i}c) = \gamma\left(\frac{1}{c}\mathbf{f}\cdot\mathbf{u}, -\mathbf{f}\right),$$

while the second follows from (35.12) by 'lowering the indices'. Comparison of these expressions shows that

$$f_i = -q\phi_{,i}, \tag{36.2}$$

as asserted. But the general case is the most important since it applies, in particular, to the motion of a charged particle in an electromagnetic field (as we shall see in Section 38 below).

We shall seek to derive the equations of motion of the particle, which in the present case are [cf. (35.12) (i)]

$$m_0 g_{\mu\nu}\frac{d^2x^\mu}{d\tau^2} = \frac{q}{c}(\Phi_{\mu,\nu} - \Phi_{\nu,\mu})U^\mu, \tag{36.3}$$

from a variational principle analogous to the classical *Hamilton principle*

$$\delta\int_{t_1}^{t_2} L dt = 0. \tag{36.4}$$

This asserts that the time-integral of the *Lagrangian function L* is stationary for the actual path of the particle. Guided by the form which the classical principle takes in the particular case (36.2), namely

$$\delta\int_{t_1}^{t_2} (\tfrac{1}{2}m_0 u^2 - q\phi)dt = 0,$$

and by the relativistic requirement of invariance, we now propose to investigate the equation

$$\delta\int_{\mathscr{P}}^{\mathscr{Q}} \left(-c^2 m_0 d\tau - \frac{q}{c}\Phi_\mu dx^\mu\right) = 0, \tag{36.5}$$

where the integration is to be performed along timelike worldlines joining two fixed events $\mathscr{P}$ and $\mathscr{Q}$ at which the particle is present. Since $-c^2 m_0 d\tau = -c^2 m_0 (1 - u^2/c^2)^{1/2} dt = (-c^2 m_0 + \frac{1}{2} m_0 u^2 + \ldots)dt$, this term of our integrand corresponds closely to its classical counterpart, aside from the constant $-c^2 m_0$ which does not affect the paths. And for the particular case (36.2) the second term of our integrand is exactly equal to the classical term $-q\phi dt$. Moreover, in the case of zero force, (36.5) asserts

$$\delta \int_{\mathscr{P}}^{\mathscr{Q}} d\tau = 0,$$

which we know to be true: time dilation makes all clocks present at $\mathscr{P}$ and $\mathscr{Q}$ indicate *shorter* proper time intervals than the clock which is free and is present at $\mathscr{P}$ and $\mathscr{Q}$ (recall the twin 'paradox').

Now introduce a monotonic parameter $\theta = \theta(\tau)$ along the paths of integration so that $\theta = \theta_1$ at $\mathscr{P}$ and $\theta = \theta_2$ at $\mathscr{Q}$. Then (36.5) can be written

$$\delta \int_{\theta_1}^{\theta_2} \Lambda d\theta = 0, \quad \Lambda = -cm_0 \sqrt{(g_{\mu\nu} \dot{x}^\mu \dot{x}^\nu)} - \frac{q}{c} \Phi_\mu \dot{x}^\mu, \quad (36.6)$$

the dot in this case denoting differentiation with respect to θ. It is known from the calculus of variations that the extremals satisfy the *Euler–Lagrange equations*

$$\frac{d}{d\theta} \frac{\partial \Lambda}{\partial \dot{x}^\nu} - \frac{\partial \Lambda}{\partial x^\nu} = 0. \quad (36.7)$$

The first term of Λ does not involve x^ν, and $\partial/\partial \dot{x}^\nu$ of this term is $-cm_0 g_{\mu\nu} \dot{x}^\mu / C$, $C = \sqrt{(g_{\mu\nu} \dot{x}^\mu \dot{x}^\nu)}$. Consequently equation (36.7) becomes

$$-cm_0 g_{\mu\nu} \frac{C\ddot{x}^\mu - \dot{x}^\mu dC/d\theta}{C^2} - \frac{q}{c} (\Phi_{\nu,\mu} \dot{x}^\mu - \Phi_{\mu,\nu} \dot{x}^\mu) = 0. \quad (36.8)$$

Now we can set $\theta = \tau$. Then $C \equiv c$ and (36.8) reduces to exactly the equation (36.3). Consequently (36.5) is the desired variational principle, though it is by no means the only possible one. If we set $\theta = t$ in (36.6) and write L for Λ, (36.6) (i) takes the form (36.4) with

$$L = -c^2 m_0 \left(1 - \frac{u^2}{c^2}\right)^{1/2} + q\mathbf{w} \cdot \mathbf{u} - q\phi, \quad \Phi_\mu = (\phi, -c\mathbf{w}). \quad (36.9)$$

This is the *relativistic Lagrangian*. The *generalized relativistic momentum* $\tilde{\mathbf{p}}$ of the particle is defined by $\tilde{\mathbf{p}} = \partial L/\partial \mathbf{u}$ (i.e. $\tilde{p}_i = \partial L/\partial u^i$) and thus, from (36.9),

$$\tilde{\mathbf{p}} = \frac{m_0 \mathbf{u}}{(1 - u^2/c^2)^{1/2}} + q\mathbf{w} = \mathbf{p} + q\mathbf{w}, \qquad (36.10)$$

where $\mathbf{p}$ is the relativistic momentum. The *Hamiltonian H* is defined by the equation $H = \mathbf{u} \cdot \partial L/\partial \mathbf{u} - L$, which, from (36.9) and (36.10), is found to be

$$H = \frac{m_0 c^2}{(1 - u^2/c^2)^{1/2}} + q\phi. \qquad (36.11)$$

But to obtain the 'canonical' form of H we must yet express it in terms of $\tilde{\mathbf{p}}$. From (36.10) and (36.11) we get

$$\left(\frac{H - q\phi}{c}\right)^2 - \left(\tilde{\mathbf{p}} - q\mathbf{w}\right)^2 = c^2 m_0^2, \qquad (36.12)$$

and so

$$H = \sqrt{[c^4 m_0^2 + c^2 (\tilde{\mathbf{p}} - q\mathbf{w})^2]} + q\phi. \qquad (36.13)$$

This is the starting point of several relativistic quantum theories, most notably that of Dirac.

Finally we write down the *Hamilton–Jacobi equation*. It is obtained from the equation for the Hamiltonian, (36.12), by replacing $\tilde{\mathbf{p}}$ by grad S and H by $-\partial S/\partial t$:

$$(\operatorname{grad} S - q\mathbf{w})^2 - \frac{1}{c^2}\left(\frac{\partial S}{\partial t} + q\phi\right)^2 + m_0^2 c^2 = 0. \qquad (36.14)$$

Here S stands for the *action*, i.e. the integral in (36.5).

Exercises V

1. How fast must a particle move before its kinetic energy equals its rest energy? $[0.866\,c]$

2. How fast must a 1 kg cannon ball move to have the same kinetic energy as a cosmic-ray proton moving with γ-factor 10^{11}? $[\sim 5\,\mathrm{ms}^{-1}]$

3. The mass of a hydrogen atom is 1.008 14 amu, that of a neutron is 1.008 98 amu, and that of a helium atom (two hydrogen atoms and two neutrons) is 4.003 88 amu. Find the binding energy as a fraction of the total energy of a helium atom.

4. A particle with four-momentum $\mathbf{P}$ is observed by an observer who moves with four-velocity $\mathbf{U}_0$. Prove that the energy of the particle relative to that observer is $\mathbf{U}_0 \cdot \mathbf{P}$.

5. A rocket propels itself rectilinearly by giving portions of its mass a constant (backward) velocity U relative to its instantaneous rest frame. It continues to do so until it attains a velocity V relative to its initial rest frame. Prove that the ratio of the initial to the final rest mass of the rocket is given by

$$\frac{M_i}{M_f} = \left(\frac{c+V}{c-V}\right)^{\frac{c}{2U}}.$$

Note that the least expenditure of mass needed to attain a given velocity occurs when $U = c$, i.e. when the rocket propels itself with a jet of photons. [*Hint*: $(-\mathrm{d}M)\,U = M\mathrm{d}u'$, where M is the rest mass of the rocket, and u' is its velocity relative to its instantaneous rest frame. Cf. also the first in the line of equations before (14.1).]

By reference to Exercise II (14), prove that, if the rocket moves with constant proper acceleration α for a proper time interval τ, then $M_i/M_f = \exp(\alpha\tau/U)$. If $U = c$, $\alpha = g$, and $\tau = n$ years, prove $M_i/M_f \approx \mathrm{e}^n$. [*Hint*: Exercise II (15).]

6. Two particles with rest masses m_1 and m_2 move collinearly in some inertial frame, with uniform velocities u_1 and u_2, respectively. They collide and form a single particle with rest mass m moving at velocity u. Prove that

$$m^2 = m_1^2 + m_2^2 + 2m_1 m_2 \gamma(u_1)\,\gamma(u_2)\,(1 - u_1 u_2/c^2),$$

and also find u. [*Hint*: for the first part, use a four-vector argument, *or* a result of Section 30.]

7. Consider a *head-on* elastic collision of a 'bullet' of rest mass M with a stationary 'target' of rest mass m. Prove that the post-collision γ-factor of the bullet cannot exceed $(m^2 + M^2)/2mM$. This means that for large bullet energies (with γ-factors much larger than this critical value), almost the entire energy of the bullet is transferred to the target. [*Hint*: if $\mathbf{P}, \mathbf{P}'$ are the pre- and post-collision four-momenta of the bullet, and $\mathbf{Q}, \mathbf{Q}'$ those of the target, show, by going to the CM frame, that $(\mathbf{P}' - \mathbf{Q})^2 \geq 0$; in fact, in the CM frame $\mathbf{P}' - \mathbf{Q}$ has no spatial components.] The situation is radically different in Newtonian mechanics, where the pre- and post-collision velocities of the bullet are related by $u/u' = (M+m)/(M-m)$. Prove this.

8. The position vector of the centre of mass of a system of particles in any inertial frame is defined by $\mathbf{r}_{CM} = \Sigma m\mathbf{r}/\Sigma m$. If the particles

suffer only collision forces, prove that $\dot{\mathbf{r}}_{CM} = \mathbf{u}_{CM}$ ($\cdot \equiv d/dt$); i.e. the centre of mass moves with the velocity of the CM frame. [*Hint*: Σm, $\Sigma m\dot{\mathbf{r}}$ are constant; $\Sigma \dot{m}\mathbf{r}$ is zero between collisions, and *at* any collision we can factor out the $\mathbf{r}$ of the participating particles: $\mathbf{r} \Sigma \dot{m} = 0$.]

9. Generalize equation (31.5), for the efficiency of an elastic bombardment, to the case where the target has rest mass N, instead of M like the bullet. Then note that for sufficiently large N, k can be arbitrarily close to unity. [*Answer*: $k^{-1} = 1 + (m + 2M)/2N$.]

10. Show that a photon cannot spontaneously disintegrate into an electron–positron pair. [*Hint*: four-momentum conservation.] But in the presence of a stationary nucleus (acting as a kind of catalyst) it can. If the rest mass of the nucleus is N, and that of the electron (and positron) is m, what is the threshold frequency of the photon? Verify that for large N the efficiency is ~ 100 per cent (cf. the preceding problem), so that the nucleus then comes close to being a pure catalyst.

11. If one neutron and one pi-meson are to emerge from the collision of a photon with a stationary proton, find the threshold frequency of the photon in terms of the rest mass n of a proton or neutron (here assumed equal) and that, m, of a pi-meson. [*Answer*: $c^2 (m^2 + 2mn)/2hn$.]

12. A fast electron of rest mass m decelerates in a collision with a heavy nucleus and emits a (bremsstrahlung) photon. Prove that the energy of the photon can range all the way up to $(\gamma - 1)mc^2$, the kinetic energy of the electron. [*Hint*: use a four-vector argument.]

13. A particle of rest mass m decays from rest into a particle of rest mass m' and a photon. Find the separate energies of these end products. [*Answer*: $c^2 (m^2 \pm m'^2)/2m$. *Hint*: use a four-vector argument.]

14. An excited atom, of total mass m, is at rest in a given frame. It emits a photon and thereby loses internal (i.e. rest) energy ΔE. Calculate the exact frequency of the photon, making due allowance for the recoil of the atom. [*Answer*: $(\Delta E/h) (1 - \frac{1}{2}\Delta E/mc^2)$. *Hint*: use a four-vector argument.]

15. A rocket propels itself rectilinearly by emitting *radiation* in the direction opposite to its motion. If V is its final velocity relative to its initial rest frame, prove *ab initio* that the ratio of the initial to the final rest mass of the rocket is given by

$$\frac{M_i}{M_f} = \left(\frac{c + V}{c - V}\right)^{1/2},$$

and compare this with the result of Exercise 5 above. [*Hint*: equate energies and momenta at the beginning and at the end of the acceleration, writing $\Sigma h\nu$ and $\Sigma h\nu/c$ for the total energy and momentum, respectively, of the emitted photons.]

16. If a photon with four-momentum **P** is observed by two observers having four-velocities $\mathbf{U}_0$ and $\mathbf{U}_1$, prove that the observed frequencies are in the ratio $\mathbf{U}_0 \cdot \mathbf{P}/\mathbf{U}_1 \cdot \mathbf{P}$. Hence rederive equation (17.3).

17. In an inertial frame S, two photons of frequencies ν_1 and ν_2 travel in the positive and negative x-directions respectively. Find the velocity of the CM frame of these photons. [*Answer*: $v/c = (\nu_1 - \nu_2)/(\nu_1 + \nu_2)$.]

18. For the 'Compton collision' discussed in Section 33 prove the relation

$$\tan \phi = (1 + h\nu/mc^2)^{-1} \cot \tfrac{1}{2}\theta.$$

19. Obtain equation (33.7), governing the head-on *inverse* Compton effect, by transforming the Compton formula (33.6) to a relevant frame with the use of (17.3).

20. Taking $h = 6.63 \times 10^{-27}$ ergs and $c = 300\,000 \,\mathrm{km\,s^{-1}}$, calculate how many photons of wavelength 5×10^{-5} cm (in the yellow-green of the visible spectrum) must fall per second on a blackened plate to produce a force of one dyne. [*Answer*: 7.3×10^{21}. *Hint*: force equals momentum absorbed per unit time.]

21. Uniform parallel radiation is observed in two arbitrary inertial frames S and S' in which it has frequencies ν and ν' respectively. If p, g, σ denote, respectively, the radiation pressure, momentum density, and energy density of the radiation in S, and primed symbols denote corresponding quantities in S', prove $p'/p = g'/g = \sigma'/\sigma = \nu'^2/\nu^2$. [*Hint*: Exercise III(17).]

22. Planck's constant h has the dimensions of action (energy $\times$ time or momentum $\times$ distance) which suggests that the action of any periodic phenomenon may have to be a multiple of h. Accordingly Bohr constructed a model of the hydrogen atom in which the action of the single orbiting electron was quantized, requiring $2\pi rmv = nh$, $n = 1, 2, \ldots$, where m is the mass of the electron, v its speed, and r the radius of the orbit. This led to a hydrogen spectrum which fitted the then known facts. Show that Bohr's hypothesis (1913) is equivalent to the assumption that a permissible orbit must contain an integral number of de Broglie electron waves.

23. If $\Delta x^{\mu} = (c\Delta t, \Delta \mathbf{r})$ is the four-vector join of two events on the worldline of a uniformly moving particle (or photon), prove that the frequency vector of its de Broglie wave is given by

$$N^{\mu} = \frac{v}{c\Delta t} \Delta x^{\mu},$$

whence $v/\Delta t$ is invariant. Compare with Exercise III (7).

24. Two identical particles move with velocities $\pm u$ along the parallel lines $z = 0, y = \pm a$ in a frame S, passing $x = 0$ simultaneously. Prove that all centroids determined by observers moving collinearly with these particles lie on the open line-segment $x = z = 0$, $|y| < ua/c$. Compare this with the dimensions of the 'minimum region' given in the note at the end of Section 34 on p. 101. Also note that, for the *same* total (relativistic) mass and angular momentum, the two particles cannot move along lines closer than $2ua/c$ without breaking the relativistic speed limit.

25. For motion under a rest mass preserving inverse square force $\mathbf{f} = -k\mathbf{r}/r^3$ ($k = $ constant), derive the energy equation $m_0 \gamma c^2 - k/r$ = constant. [*Hint*: Equation (35.8).]

26. Prove that, in relativistic as in Newtonian mechanics, the time rate of change of the angular momentum $\mathbf{h} = \mathbf{r} \times \mathbf{p}$ of a particle about an origin O is equal to the couple $\mathbf{r} \times \mathbf{f}$ of the applied force about O. If $L^{\mu\nu}$ is the particle's four-angular momentum, and if we define the *four-couple* $G^{\mu\nu} = x^{\mu} F^{\nu} - x^{\nu} F^{\mu}$, where x^{μ} and F^{μ} are the four-vectors corresponding to $\mathbf{r}$ and $\mathbf{f}$, prove that $(\mathrm{d}/\mathrm{d}\tau) L^{\mu\nu} = G^{\mu\nu}$ and that the space–space part of this equation corresponds to the above three-vector result.

27. A particle moves rectilinearly under a rest mass preserving force in some inertial frame. Show that the product of its rest mass and its instantaneous proper acceleration equals the magnitude of the relativistic three-force acting on the particle in that frame. [*Hint*: (14.1) and (35.14).] Show also that this is not necessarily true when the motion is not rectilinear.

28. In an inertial frame there is a uniform electric field $\mathbf{e}$ in the positive x-direction, which exerts a rest mass preserving force $\mathbf{f} = q\mathbf{e}$ on a particle carrying a charge q. If the particle is released from rest at the origin, and m, τ denote its rest mass and proper time, prove that the particle will move (*hyperbolically*) according to the equation

$$x = (c/Q)(\cosh Q\tau - 1), \quad Q = qe/cm_0.$$

VI
RELATIVITY AND ELECTROMAGNETISM

37. Introduction

Having examined and relativistically modified Newtonian particle mechanics, it would be natural to look next with the same intentions at Maxwell's electrodynamics. But that theory turns out to be already 'special-relativistic'. In other words, its basic laws, as summarized by the four Maxwell equations plus Lorentz's force law, are form-invariant under Lorentz transformations, i.e. under transformations from one inertial frame to another. Indeed it was the problem of finding a transformation that leaves Maxwell's equations invariant which led Lorentz to the discovery (1895) of the equations now associated with his name (without, of course, at that time fully appreciating their physical significance).

However, though relativity did not *modify* Maxwell's theory in vacuum,[1] it added immeasurably to our understanding of it, and also gave us new tools for working in it: the four-tensor calculus and the method of 'frame hopping'. It was relativity that brought to light the basic formal simplicity of Maxwell's theory. Within its old Galilean framework, that theory seemed quite far-fetched. Within relativity, on the other hand, it is one of the two or three simplest possible theories of a field of force. (Other candidates for this distinction are probably the Yukawa–Klein–Gordon scalar meson field theory, and Nordström's attempt at a special-relativistic theory of gravity.) Any clever mathematician, asked to produce formal Lorentz-invariant field theories without regard to empirical data would probably come up with those three in short order. That two of them apply very accurately to certain natural phenomena is really quite remarkable.

In fact, rather than *verify* that Maxwell's equations are Lorentz-invariant, we shall use just such a synthetic approach, which will parallel what we did in mechanics, and bring out once more the 'man-made' aspect of physical laws. We shall construct a relativistic field theory that is consistent with just a very few of the basic facts of electromagnetism, and we shall find that it is Maxwell's.[2]

¹ On the other hand, the theory of electromagnetism in moving media, originally due to Minkowski (1908), is a purely relativistic development. But we shall not enter into that.

² The two key ideas of this construction I owe to Ivor Robinson.

38.   The formal structure of Maxwell's theory

The only two assumptions that we shall specifically make about the electromagnetic force will be that it is a *pure* force (i.e. rest mass preserving) and that it acts on particles in proportion to the charge q which they carry. Beyond that, only 'simplicity' and some analogies with Newton's gravitational theory will guide us. We begin by considering—and rejecting—certain simple possibilities. Consider a field of three-force **f** which, like the Newtonian gravitational force, acts on a particle independently of its velocity, in some frame S. From the transformation equations (35.6) for **f** we then see that in another frame S' such a force *will* depend on the velocity **u** of the particle on which it acts. So velocity-independence is not a Lorentz-invariant condition we can impose on a field of three-force.

We could, however, suppose that there exists a field of *four*-force **F** which acts on any particle independently of its four-velocity **U** [such as the gradient field (35.10)]. By (35.5), the corresponding three-force would be $\mathbf{f} = \gamma^{-1}(u)(F^1, F^2, F^3)$, which, as expected, depends on the particle's velocity. But it would not be a *pure* force [cf. (35.9)], since, for fixed **F**, $\mathbf{U} \cdot \mathbf{F}$ could not vanish for arbitrary **U** unless **F** itself vanished. So we must here reject this type of field.

The next simplest case, and the one that actually applies in Maxwell's theory, is that of a force which everywhere depends *linearly* on the velocity of the particles on which it acts. In special relativity it is natural to make this requirement on the respective four-vectors **F** and **U**:

$$F_\mu = \frac{q}{c} E_{\mu\nu} U^\nu, \tag{38.1}$$

where the 'coefficients' $E_{\mu\nu}$ in this linear relation must be tensorial to make it Lorentz-invariant (quotient rule!) if—as is natural—we take the charge q of the particle in question to be a scalar invariant. (The inclusion of c is for later convenience.) In this theory no 'moving' charge (analogous to the 'moving' mass $m = \gamma m_0$) arises. We regard $E_{\mu\nu}$ as the *field tensor* which can be determined in practice by use of *test* charges (so as not to alter it) and (38.1). Whereas in Newtonian gravitational theory, for example, the *field* **g** and the *force* **f** = *m***g** are very closely related and in fact often confused, in our present theory the distinction between field and force becomes very apparent.

If we next demand that the force (38.1) be pure, we need

$$F_\mu U^\mu = (q/c)E_{\mu\nu}U^\mu U^\nu = 0,$$

for all U^μ, and hence

$$E_{\mu\nu} = -E_{\nu\mu}, \tag{38.2}$$

i.e. the field tensor must be *antisymmetric*. (It is easy to see this by giving U^μ only a 0th component, then a 0th and 1st, 0th and 2nd, etc.)

The field affects the charges in accordance with equation (38.1). How do the charges, reciprocally, affect the field? The answer is given by the *field equations*. One of the characteristics of a field theory is that the action of the sources can (though it need not) spread through the field at a finite speed, and it is the *field* which eventually acts on the test particles, rather than the sources themselves by some 'action at a distance'. The field equations are therefore *differential* equations, telling how the sources affect the field in their vicinity, and how one part of the field affects a neighbouring part. Newton's theory, too, can be regarded as a field theory (where, however, effects spread instantaneously through the field), with field equation div **g** = $-4\pi G\rho$, ρ being the mass density and G the constant of gravity.

In our present theory the analogue of div **g** ($g^i{}_{,i}$) is $E^{\mu\nu}{}_{,\mu}$. Since the field is a tensor (one more index than in the Newtonian theory), the source must be a vector. Now the field affects the charges according to their velocities; it would therefore be quite reasonable to suppose that, reciprocally, the charges affect the field also according to their velocities. Hence we are led to postulate the following field equations:

$$E^{\mu\nu}{}_{,\mu} = k\rho_0 U^\nu = kJ^\nu, \tag{38.3}$$

where k is some universal constant, U^μ the four-velocity of the source distribution, ρ_0 its *proper charge density* (i.e. the charge density as measured in the rest frame of the sources—a scalar), and where J^μ is defined by this equation: it is called the *four-current density*. In standard terminology, what we envisage our source to be is a *convection current*. (If the elementary charges move incoherently, the 'rest frame' is that for which, in the limit of small volumes, there is no *net* current, i.e. $\Sigma q\mathbf{u} = 0$; and U^μ is *its* four-velocity.)

A significant consequence of the field equations (38.3) is the so-called *equation of continuity*

$$J^{\nu}{}_{,\nu} = 0, \tag{38.4}$$

which we shall presently interpret as the *conservation of charge*. We obtain it from (38.3) by use of the antisymmetry of $E^{\mu\nu}$ and the symmetry of second derivatives:

$$kJ^{\nu}{}_{,\nu} = E^{\mu\nu}{}_{,\mu\nu} = E^{\nu\mu}{}_{,\nu\mu} = -E^{\mu\nu}{}_{,\mu\nu}.$$

This result is clearly a recommendation for these field equations. Nevertheless the field equations (38.3), by themselves, are insufficient to determine $E^{\mu\nu}$: for there are *six* independent components of an antisymmetric tensor like $E^{\mu\nu}$ [cf. Exercise IV(6)], and we have only *four* field equations. This is quite similar to the Newtonian case, where we need to determine the *three* components of **g** but so far have only *one* field equation. The remedy in both cases is the same. One assumes (if one follows the present synthetic route) the existence of a *potential*. In the Newtonian case, we put $g_i = -\phi_{,i}$, and then the field equation becomes $\Sigma\phi_{,ii} = 4\pi G\rho$; this gives ϕ, which in turn gives **g**. In the field theory we are presently constructing the potential has to be a vector, Φ_{μ}, called the *four-potential*, and it determines the (antisymmetric) field $E_{\mu\nu}$ through the equation[1]

$$E_{\mu\nu} = \Phi_{\nu,\mu} - \Phi_{\mu,\nu}. \tag{38.5}$$

Then the field equations (38.3) read (after we lower ν):

$$g^{\mu\sigma} (\Phi_{\nu,\sigma\mu} - \Phi_{\sigma,\nu\mu}) = kJ_{\nu}. \tag{38.6}$$

These four equations are just enough to determine Φ_{μ}, which in turn determines $E_{\mu\nu}$.

A necessary consequence of (38.5) is the relation

$$E_{\mu\nu,\sigma} + E_{\nu\sigma,\mu} + E_{\sigma\mu,\nu} = 0, \tag{38.7}$$

as can be verified at once. In the theory of differential equations the converse is also well known: (38.7) is not only a necessary but also a sufficient condition for a four-potential Φ_{μ} satisfying (38.5) to exist.[2] [Compare the analogous condition $g_{i,j} - g_{j,i} = 0$ (curl **g** = 0) for the existence of a *scalar* potential ϕ such that $g_i = -\phi_{,i}$ (**g** = $-$ grad ϕ).] So we can regard equation (38.7) as equivalent to the hypothesis (38.5), and consequently as a *second field equation*, which, together with (38.3), uniquely determines the field. [In the Newtonian case, the two corresponding field equations are $\operatorname{div}\mathbf{g} = -4\pi G\rho$ and $\operatorname{curl}\mathbf{g} = 0$.]

Different potentials Φ_μ and $\mathring{\Phi}_\mu$ can give rise to the same field $E_{\mu\nu}$. If so, the vector $\Psi_\mu = \Phi_\mu - \mathring{\Phi}_\mu$ must satisfy $\Psi_{\mu,\nu} - \Psi_{\nu,\mu} = 0$, which implies that $\Psi_\mu = \Psi_{,\mu}$, i.e.

$$\Phi_\mu = \mathring{\Phi}_\mu + \Psi_{,\mu} \tag{38.8}$$

for some scalar Ψ. So any two four-potentials differ by a gradient. Now for a given $\mathring{\Phi}_\mu$, this gradient—the so-called *gauge*—can be chosen so as to make

$$\Phi^\mu_{,\mu} = 0. \tag{38.9}$$

It is merely necessary to satisfy

$$g^{\mu\nu}\mathring{\Phi}_{\mu,\nu} + g^{\mu\nu}\Psi_{,\mu\nu} = 0, \quad \text{i.e.} \quad \Box\Psi = -g^{\mu\nu}\mathring{\Phi}_{\mu,\nu}, \tag{38.10}$$

where

$$\Box \equiv {}_{,\mu\nu}g^{\mu\nu} \equiv \frac{1}{c^2}\frac{\partial^2}{\partial t^2} - \frac{\partial^2}{\partial x^2} - \frac{\partial^2}{\partial y^2} - \frac{\partial^2}{\partial z^2}. \tag{38.11}$$

In the theory of differential equations this equation is known to be in general solvable for Ψ, relative to any inertial frame S. In fact, it can be shown that

$$\Psi(P) = \frac{1}{4\pi}\iiint \frac{[F]\,dV}{r} \Rightarrow \Box\Psi = F, \tag{38.12}$$

where $F = F(ct, x, y, z)$ is any integrable function (it must be 'sufficiently small' at infinity), $[F]$ denotes the value of F 'retarded' by the light-travel time to the origin P of the position vector $\mathbf{r}$ of dV, and the volume integral extends over the entire three-space in S. We can therefore assume, without loss of generality, that the potential satisfies the '*Lorentz gauge condition*' (38.9), and then the field equations (38.6) decouple and simplify to

$$\Box\Phi_\mu = kJ_\mu, \tag{38.13}$$

in conjunction with (38.9).

Note that in charge-free regions ($J_\mu = 0$), equation (38.13) reduces to the wave equation with speed c, showing that disturbances of the potential in vacuum are propagated at the speed of light. The potential, however, is often regarded as an 'unphysical' auxiliary. But we find at once from (38.5), (38.13), and the commutativity of partial derivatives, that when $J_\mu = 0$ the field $E_{\mu\nu}$ *itself* satisfies the wave equation

$$\Box E_{\mu\nu} = 0. \tag{38.14}$$

Hence in the present theory disturbances of the *field* propagate at the speed of light.

Maxwell's theory now stands in essence before us. To make the connection with its usual three-dimensional formulation in terms of the *electric field* **e** and the *magnetic field* **b** (both are here written with small letters to conform to our convention for three-vectors), we *define* **e** and **b** in each inertial frame by the first of the following equations,

$$E_{\mu\nu} = \begin{pmatrix} 0 & e_1 & e_2 & e_3 \\ -e_1 & 0 & -cb_3 & cb_2 \\ -e_2 & cb_3 & 0 & -cb_1 \\ -e_3 & -cb_2 & cb_1 & 0 \end{pmatrix},$$

$$E^{\mu\nu} = \begin{pmatrix} 0 & -e_1 & -e_2 & -e_3 \\ e_1 & 0 & -cb_3 & cb_2 \\ e_2 & cb_3 & 0 & -cb_1 \\ e_3 & -cb_2 & cb_1 & 0 \end{pmatrix}, \tag{38.15}$$

while the second results from the first by raising the indices. (Bowing to the pressure of convention we here adopt SI units, though they are distinctly inconvenient for theoretical work, and blur the inherent symmetry between **e** and **b**. In Gaussian units we would write **b** for c**b**.) It was shown in Exercise IV(7) that the triplets (e_1, e_2, e_3) and (b_1, b_2, b_3) defined by (38.15) indeed *do* behave as three-vectors under spatial rotations, although this will be apparent also from the formulae we derive below, especially (38.23).

Our starting point, the four-force law (38.1), now takes the familiar form of *Lorentz's force law*

$$\mathbf{f} = q\,(\mathbf{e} + \mathbf{u} \times \mathbf{b}), \tag{38.16}$$

when we substitute into it the expressions $F_\mu = \gamma\,(\mathbf{f}\cdot\mathbf{u}/c,\ -\mathbf{f})$ [cf. (35.12)] and $U^\mu = \gamma\,(c, \mathbf{u})$, and equate the spatial components; equating the temporal component yields a mere corollary of (38.16), $\mathbf{f}\cdot\mathbf{u} = q\mathbf{e}\cdot\mathbf{u}$. We may note parenthetically that *any* pure three-force that is velocity-independent in one frame S must be a Lorentz-type force. For the corresponding four-fource can then be expressed as in

(38.1) with $E_{\mu\nu}$ defined by its components in S, namely an array like (38.15)(i) with zero bs. In other frames there will be non-zero bs.

Next we elaborate the definition (38.3) of J^μ as follows:

$$J^\mu = \rho_0 U^\mu = \rho_0 \gamma (c, \mathbf{u}) = (c\rho, \mathbf{j}), \tag{38.17}$$

where we have introduced the *charge density* ρ and the *three-current density* $\mathbf{j}$:

$$\rho = \rho_0 \gamma, \quad \mathbf{j} = \rho \mathbf{u}. \tag{38.18}$$

Evidently ρ is the charge per unit volume in the frame of observation, since in that frame a unit proper volume has measure $1/\gamma$, by length contraction. And $\mathbf{j}$ is the amount of charge crossing a fixed unit area at right angles to its motion, per unit time. The equation of continuity (38.4) then takes the form

$$\frac{\partial \rho}{\partial t} + \operatorname{div} \mathbf{j} = 0, \tag{38.19}$$

whose meaning is clear: since $\operatorname{div} \mathbf{j}$ measures the outflux of charge from a unit volume in unit time, it says that to the precise extent that charge leaves a small region, the total charge inside that region must decrease. This is an expression of *charge conservation*.

We are now ready to 'translate' the field equations (38.3) and (38.7). The SI choice for k is $1/c\varepsilon_0$. (In Gaussian units we would set $k = 4\pi/c$, i.e. $\varepsilon_0 = 1/4\pi$.) Giving ν in (38.3) the values 0, 1, 2, 3 in turn, we find that that equation is equivalent to the following two three-vector equations

$$\operatorname{div} \mathbf{e} = \frac{1}{\varepsilon_0} \rho, \quad \operatorname{curl} \mathbf{b} = \frac{1}{c^2 \varepsilon_0} \mathbf{j} + \frac{1}{c^2} \frac{\partial \mathbf{e}}{\partial t}, \tag{38.20}$$

which are two of the familiar *Maxwell equations*. The other two Maxwell equations are the three-vector equivalents of the 'potential condition' (38.7). Assigning values 1, 2, 3; 2, 3, 0; 3, 0, 1; 0, 1, 2 in turn to the indices μ, ν, σ in (38.7) (all other combinations either reproduce one of the equations so obtained or $0 = 0$), we find indeed

$$\operatorname{div} \mathbf{b} = 0, \quad \operatorname{curl} \mathbf{e} = -\frac{\partial \mathbf{b}}{\partial t}. \tag{38.21}$$

Finally we define a three-scalar potential φ and a three-vector potential $\mathbf{w}$ in each inertial frame by the equation

$$\Phi_\mu = (\varphi, -c\mathbf{w}). \tag{38.22}$$

(In Gaussian units we would write **w** for c**w**.) Then the relation (38.5) between the field and potential translates into

$$\mathbf{e} = -\operatorname{grad} \varphi - \frac{\partial \mathbf{w}}{\partial t}, \quad \mathbf{b} = \operatorname{curl} \mathbf{w}; \tag{38.23}$$

the gauge condition (38.9) translates into

$$\frac{\partial \varphi}{\partial t} + c^2 \operatorname{div} \mathbf{w} = 0; \tag{38.24}$$

and the field equations (38.13) for the potential translate into

$$\Box \, \varphi = \frac{1}{\varepsilon_0} \rho, \quad \Box \mathbf{w} = \frac{1}{c^2 \varepsilon_0} \mathbf{j}. \tag{38.25}$$

The remainder of Maxwell's theory in vacuum 'merely' consists in working out the implications of the basic equations we have developed here. Some of this will be done in the next few sections. *The interested reader may wish to bear in mind that in order to change any of our equations to Gaussian form he need only replace ε_0 by $1/4\pi$, **b** by **b**$/c$, and **w** by **w**$/c$.*

---

[1] The reader should be warned that, as regards the *four*-dimensional formulation of Maxwell's theory, conflicting conventions are used in the literature. Thus, some authors write $F_v = (q/c)E_{\mu v}U^\mu$ instead of (38.1), or $E^{\mu v},_v = kJ^\mu$ instead of (38.3), or $E_{\mu v} = \Phi_{\mu,v} - \Phi_{v,\mu}$ instead of (38.5). There is a further ambiguity in the definition of $\overset{*}{E}_{\mu v}$ [see (39.3)], depending on whether ct is taken as first or last coordinate, and accordingly on whether e_{0123} or e_{1234} is taken to be 1.

[2] See, for example, Synge, J. L. (1956) *Relativity: The Special Theory*, North Holland, Amsterdam, p.345.

39. Transformation of e and b. The dual field

From the tensor character of either of the arrays in (38.15) we can at once deduce the transformation laws of **e** and **b**. For example, if we apply the typical transformation (22.8) to $E^{\mu v}$, we find

$$-cb'_3 = \gamma(-cb_3 + ve_2/c).$$

In the same way we can obtain all the other entries in the following list:

$$e'_1 = e_1, \qquad e'_2 = \gamma(e_2 - vb_3), \qquad e'_3 = \gamma(e_3 + vb_2) \tag{39.1}$$

$$b'_1 = b_1, \qquad b'_2 = \gamma(b_2 + ve_3/c^2), \qquad b'_3 = \gamma(b_3 - ve_2/c^2). \tag{39.2}$$

Thus on transforming from one frame to another, the $\mathbf{e}$ and $\mathbf{b}$ fields get thoroughly mingled. That was, of course, immediately to be expected from the way they both enter into the formation of a single tensor, $E^{\mu\nu}$. For example, a field which is either purely electric or purely magnetic in one frame, will have both electric and magnetic components in the general frame. This 'explains' the deflection of a moving charge in a purely magnetic field: in its rest frame the charge is accelerated *only* by an $\mathbf{e}$ field. A similar 'explanation' can be given for the magnetic field of a uniformly moving charge: it arises as part of the transform of the pure Coulomb field in the rest frame.

Just as there is an invariant combination of the components of a four-vector A^{μ}, namely $\mathbf{A}^2 = (A^0)^2 - (A^1)^2 - (A^2)^2 - (A^3)^2$, which remains the same in spite of the impermanence of the individual components, so there are in fact two such invariants associated with any antisymmetric tensor $E^{\mu\nu}$. For this and other purposes it is well to introduce the *dual* $B_{\mu\nu}$ of the field tensor $E_{\mu\nu}$, defined by

$$B_{\mu\nu} := \overset{*}{E}_{\mu\nu} = \tfrac{1}{2}\varepsilon_{\mu\nu\rho\sigma}E^{\rho\sigma}. \tag{39.3}$$

[The reader should consult, if necessary, Exercises A(15)–A(20) for the definition of $\varepsilon_{\mu\nu\rho\sigma}$ and related matters.] In special relativity, where the determinant g of the metric tensor $g_{\mu\nu}$ is -1, we have

$$\varepsilon_{\mu\nu\rho\sigma} = -\varepsilon^{\mu\nu\rho\sigma} = e_{\mu\nu\rho\sigma}, \tag{39.4}$$

$e_{\mu\nu\rho\sigma}$ being the four-dimensional permutation symbol such that $e_{0123} = 1$. Thus $B_{01} = E^{23}$, $B_{12} = E^{03}$, etc. In this way we find, from (38.15), the first of the following equations, while the second again results from the first by raising the indices:

$$B_{\mu\nu} = \begin{pmatrix} 0 & -cb_1 & -cb_2 & -cb_3 \\ cb_1 & 0 & -e_3 & e_2 \\ cb_2 & e_3 & 0 & -e_1 \\ cb_3 & -e_2 & e_1 & 0 \end{pmatrix},$$

$$B^{\mu\nu} = \begin{pmatrix} 0 & cb_1 & cb_2 & cb_3 \\ -cb_1 & 0 & -e_3 & e_2 \\ -cb_2 & e_3 & 0 & -e_1 \\ -cb_3 & -e_2 & e_1 & 0 \end{pmatrix}. \tag{39.5}$$

Observe that $B_{\mu\nu}$[resp. $B^{\mu\nu}$] is obtained from $E_{\mu\nu}$[resp. $E^{\mu\nu}$] by making the replacements $\mathbf{e} \rightarrow -c\mathbf{b}$, $c\mathbf{b} \rightarrow \mathbf{e}$. The two invariants of

$E^{\mu\nu}$—immediately recognizable as such from their mode of formation—can then be expressed as follows:

$$X = \tfrac{1}{2}E_{\mu\nu}E^{\mu\nu} = -\tfrac{1}{2}B_{\mu\nu}B^{\mu\nu} = c^2 b^2 - e^2, \tag{39.6}$$

$$Y = \tfrac{1}{4}B_{\mu\nu}E^{\mu\nu} = c\mathbf{b}\cdot\mathbf{e}. \tag{39.7}$$

There are no others, except, of course, combinations of these two [cf. Exercise VI(5)]. Physically they tell us, for example, that if the magnitudes of the electric and magnetic fields are 'equal' ($cb = e$) at some event in one frame, they are equal in all frames; and if these fields are orthogonal in one frame ($\mathbf{b}\cdot\mathbf{e} = 0$), they are orthogonal in all frames. If a field is purely magnetic at some event in one frame ($X > 0$), it cannot be purely electric in another, and vice versa; if the angle between $\mathbf{e}$ and $\mathbf{b}$ is acute in one frame ($Y > 0$), it cannot be obtuse in another. When $X = Y = 0$ the field is said to be *null* and $\mathbf{e}$ is perpendicular to $\mathbf{b}$ and $cb = e$ in all frames.

Can any given field be either purely electric or purely magnetic in *some* frame? Obviously not unless $Y = 0$; and if $X \geqslant 0$ the field can never be purely electric, while if $X \leqslant 0$ it can never be purely magnetic. But suppose $X > 0$ and $Y = 0$, i.e.

$$c^2 b^2 - e^2 > 0, \quad \mathbf{b}\cdot\mathbf{e} = 0. \tag{39.8}$$

Then if $\mathbf{e} = 0$ to an observer having four-velocity $U^\mu = \gamma(c, \mathbf{u})$, we have $E_{\mu\nu}U^\nu = 0$ (since this reads $E_{\mu 0} = 0$ in his rest frame).[1] In the general frame it reads [cf. (38.1) and (38.16)]

$$\mathbf{e}\cdot\mathbf{u} = 0, \quad \mathbf{e} + \mathbf{u}\times\mathbf{b} = 0. \tag{39.9}$$

We need to solve these equations for $\mathbf{u}$. By (39.8)(ii) and (39.9)(i) we can write, for some numbers λ and μ,

$$\mathbf{u} = \lambda\mathbf{b} + \mu(\mathbf{e}\times\mathbf{b}). \tag{39.10}$$

When substituted into (39.9)(ii) this yields $\lambda = $ arbitrary, $\mu = b^{-2}$. The only remaining question is whether $\mathbf{u}$ so determined satisfies the equation $\mathbf{u}^2 < c^2$, i.e. by (39.10) and (39.8)(ii),

$$\lambda^2 b^4 < c^2 b^2 - e^2.$$

But the right-hand side is positive by hypothesis, so there will indeed be a possible range for λ, and hence a set of frames in which the field is purely magnetic. Similarly, if $Y = 0$ and $X < 0$ there is an analogous set of frames in which the field is purely electric. (The relevant equation to be solved is then $B_{\mu\nu}U^\nu = 0$.) Note that all the frames determined by this analysis have a common relative direction of motion.

It can also be shown [cf. Exercise VI(13)] that when $Y \neq 0$ there are infinitely many frames (with common relative direction of motion) in which **e** is *parallel* to **b**. In fact, our above case is a limit of this latter case.

We remark finally that in terms of $B^{\mu\nu}$ the second tensor field equation, (38.7), can be written in the form

$$\tfrac{1}{2}\varepsilon^{\mu\nu\sigma\tau}E_{\mu\nu,\sigma} = B^{\sigma\tau}{}_{,\sigma} = 0. \tag{39.11}$$

So the dual field is 'source-free'. Note how (38.21) results *formally* from (38.20) as does (39.11) from (38.3): by setting the source terms equal to zero and replacing **e** by $-c$**b** and c**b** by **e**. A possible source for $B^{\mu\nu}$ would be a divergence-free *magnetic* current density J^ν_{mag}, in analogy with (38.3), and then the theory would be completely symmetric in **e** and c**b** (at the cost, however, of having no four-potential). Indeed, this is another 'natural' field theory within the framework of special relativity, of which Maxwell's is a particular case. (Though apparently there are 8 equations for 6 unknowns in this theory, these equations satisfy 2 differential identities—$J^\nu_{,\nu} = 0$ and $J^\nu_{\text{mag},\nu} = 0$—which effectively reduce the degrees of freedom to 6.) But at present there is absolutely no empirical evidence for the existence of magnetic charges ('monopoles'), in spite of Dirac's suggestion (1931) that they would automatically explain the quantization of electric charge, and of the consequent efforts of the experimentalists to find them.

[1] The following argument is taken from N. Woodhouse, mimeographed *Notes on Special Relativity*, Oxford, 1980, p. 77.

40.   Potential and field of an arbitrarily moving charge

This section will illustrate some of the power that relativity has brought to the treatment of electromagnetic theory and problems. In particular, we shall derive the potential and field of an arbitrarily moving point charge essentially by transforming the simple Coulomb potential of a stationary charge.

The Coulomb potential is derived most easily by applying the integral solution (38.12) to the field equations (38.13):

$$\Phi_\mu = (\varphi, -c\mathbf{w}) = \frac{1}{4\pi\varepsilon_0 c} \int\int\int \frac{[J_\mu]\mathrm{d}V}{r}. \tag{40.1}$$

It is shown in all the standard textbooks that this indeed satisfies (38.13) and the gauge condition (38.9) and is unique under certain reasonable conditions (no incoming radiation, no advanced potentials). In the case of a point charge q at rest at the origin of a frame S (40.1) immediately gives

$$\varphi = Q/r, \quad \mathbf{w} = 0, \quad [Q = (4\pi\varepsilon_0)^{-1} q], \tag{40.2}$$

where r is distance from the origin, and where we have introduced the symbol Q to save ballast. Now it is of particular importance for what follows to observe that even if the charge is only *momentarily* at rest at the spatial origin $(0, 0, 0)$ of S, say at $t = 0$, then the potential derived from (40.1) is still (40.2) at all events from which the event $(0, 0, 0, 0)$ is 'retarded', i.e. at all events on the forward light-cone of $(0, 0, 0, 0)$ (see Fig. 16). To see this, consider the point charge as the limit of a finite distribution of charge, and observe that the integral in (40.1) involves the velocity but not the acceleration of that distribution.

In general we have two essentially different methods for transforming a situation from one specific frame to another, or to the general frame. We can translate it directly, by use of transformation formulae, as we did for example in the problem of relativistic billiards in Section 29. Or we can try to spot a tensor equation which holds in the

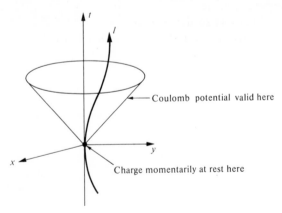

Fig. 16

specific frame (and therefore in any other) and then reinterpret it in the general frame. That is the method we shall use here.

Consider an arbitrarily moving charge q, with proper time τ, following a worldline l with equation $y^\mu = y^\mu(\tau)$. Let $\mathscr{Q}$ be a specific event on l at which q has four-velocity $U^\mu = dy^\mu/d\tau$ and four-acceleration $A^\mu = dU^\mu/d\tau$. Consider also an event $\mathscr{P}$ with coordinates x^μ on the forward light-cone of $\mathscr{Q}$, at which we wish to evaluate the potential (see Fig. 17). Let us write

$$R^\mu = x^\mu - y^\mu(\tau) = :(ct, \mathbf{r}) \qquad (40.3)$$

for the connection vector $\mathscr{Q}\mathscr{P}$. If we think of x^μ as *given*, then τ at the 'retarded' event $\mathscr{Q}$ is determined by (40.3) together with the null and future-pointing requirements on R^μ:

$$R_\mu R^\mu = 0, \quad t > 0, \quad \text{so} \quad r = ct. \qquad (40.4)$$

Now we assert (or 'spot') that the four-potential at $\mathscr{P}$ is given by:

$$\Phi_\mu = \frac{QU_\mu}{R_\sigma U^\sigma}. \qquad (40.5)$$

This tensor equation (!) can be validated at once by looking at its components in the rest frame of the charge at $\mathscr{Q}$:

$$(\Phi_\mu)_{\text{rest frame}} = \frac{Q(c, \mathbf{0})}{c^2 t} = \frac{Q(1, \mathbf{0})}{r}, \qquad (40.6)$$

where we have used the relation $r = ct$ [cf. (40.4)]. But (40.6) is just the potential (40.2), and so the general validity of (40.5) is established.

We could now proceed at once to calculate the field from (40.5)

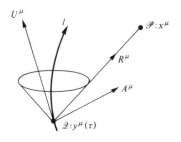

F<small>IG</small>.   17

according to (38.5). But we shall pause for a moment to rewrite (40.5) in more familiar terms. In the general frame S it reads

$$\Phi^\mu = \frac{Q\gamma(c, \mathbf{u})}{(ct, \mathbf{r}) \cdot \gamma(c, \mathbf{u})} = \frac{Q(c, \mathbf{u})}{cr - \mathbf{r} \cdot \mathbf{u}}, \tag{40.7}$$

where again we have used $r = ct$. But $-\mathbf{r} \cdot \mathbf{u}/r$ is now u_r, the radial velocity of the source away from the observation point in S. Hence (40.7) can be written in the form

$$\varphi = \frac{Q}{[r(1 + u_r/c)]}, \quad \mathbf{w} = \frac{Q}{c^2}\left[\frac{\mathbf{u}}{r(1 + u_r/c)}\right], \tag{40.8}$$

where the square brackets remind us to evaluate all enclosed quantities at the retarded event. These are the well-known *Liénard–Wiechert potentials* of a moving charge. [If one *naively* evaluates them from (40.1), one might miss the u_r factor: that arises because during the sweep of the 'collection sphere' at the speed of light towards the observation point, the charge—when not regarded as a geometric point—moves, and is 'seen' for a longer or shorter time than if it were stationary, depending on the sign of u_r.]

Now back to (40.5) and to the calculation of the field. We shall find that whereas the potential is unaffected by the acceleration of the charge, the same is not true of the field, and it is this which complicates the calculation. But first we discuss some preliminaries. The partial derivatives of the potential in equation (38.5) evidently refer to the field point, x^μ. As we move that point slightly, we induce a motion of the 'retarded' event $\mathscr{Q}$ along l, i.e. a slight change in τ; we shall write τ_ν for $\partial\tau/\partial x^\nu$. Then, for quantities defined at $\mathscr{Q}$,

$$\frac{\partial}{\partial x^\nu} = \tau_\nu \frac{\mathrm{d}}{\mathrm{d}\tau}, \tag{40.9}$$

whence, in particular, from (40.3),

$$R^\mu_{,\nu} = \delta^\mu_\nu - U^\mu \tau_\nu, \tag{40.10}$$

and then, from (40.4) and (40.10),

$$R_\mu R^\mu_{,\nu} = 0 = R_\mu(\delta^\mu_\nu - U^\mu \tau_\nu).$$

This yields

$$\tau_\nu = R_\nu / R_\mu U^\mu. \tag{40.11}$$

It will also be convenient to introduce a symbol D by the first of the following equations,

$$D = R_\sigma U^\sigma, \quad D_{,\mu} = R_{\sigma, \mu} U^\sigma + R_\sigma A^\sigma \tau_\mu, \tag{40.12}$$

while the second then results from (40.9).

The actual calculation of the field $E_{\mu\nu} = \Phi_{\nu, \mu} - \Phi_{\mu, \nu}$ can now begin. From (40.5) we have, using our various preliminaries,

$$\Phi_{\nu, \mu} = \frac{Q}{D^2} [D A_\nu \tau_\mu - U_\nu (R_{\sigma, \mu} U^\sigma + R_\sigma A^\sigma \tau_\mu)].$$

Into this we substitute from (40.10), (40.11), and (40.12) to find

$$\Phi_{\nu, \mu} = -\frac{Q}{D^2} U_\nu U_\mu + \frac{Qc^2}{D^3} R_\mu \tilde{U}_\nu, \tag{40.13}$$

where

$$\tilde{U}_\nu = U_\nu - c^{-2}(U_\nu A_\sigma - U_\sigma A_\nu) R^\sigma. \tag{40.14}$$

The first term on the right-hand side of (40.13) drops out when we antisymmetrize, and so we find, for the field $E_{\mu\nu}$ of a moving charge q,

$$E_{\mu\nu} = \frac{Qc^2 (R_\mu \tilde{U}_\nu - R_\nu \tilde{U}_\mu)}{(R_\sigma U^\sigma)^3}, \quad Q = \frac{q}{4\pi\varepsilon_0}. \tag{40.15}$$

It remains to express this in three-vector form. If $\mathbf{r}$ is the connection vector from the (retarded) source point to the field point, in the inertial frame of interest, we have from (44.3) and (40.4),

$$R_\mu = (r, -\mathbf{r}). \tag{40.16}$$

Then, writing γ, $\mathbf{u}$, $\mathbf{a}$ for the Lorentz factor, velocity, and acceleration of the charge *at the retarded event*, and making use of (40.16), (23.4)(iv), and (23.5)(iv), we find from (40.14) that

$$\tilde{U}_0 = \gamma c^{-1}(c^2 + \gamma^2 \mathbf{r} \cdot \mathbf{a}),$$
$$\tilde{U}_i = -\gamma c^{-2}[(c^2 + \gamma^2 \mathbf{r} \cdot \mathbf{a})\mathbf{u} + \gamma D \mathbf{a}].$$

These values can now be substituted into (40.15) to calculate $e_i = E_{0i}$; in this way we find

$$\mathbf{e} = \frac{Q}{\gamma^2 (cr - \mathbf{r} \cdot \mathbf{u})^3} [(c^2 + \gamma^2 \mathbf{r} \cdot \mathbf{a})(c\mathbf{r} - r\mathbf{u}) - r\gamma^2 (cr - \mathbf{r} \cdot \mathbf{u})\mathbf{a}], \tag{40.17}$$

which, alternatively, can be written

$$\mathbf{e} = \frac{Q}{(cr - \mathbf{r} \cdot \mathbf{u})^3} \left[\frac{c^2}{\gamma^2}(c\mathbf{r} - r\mathbf{u}) + \mathbf{r} \times \{(c\mathbf{r} - r\mathbf{u}) \times \mathbf{a}\} \right]. \quad (40.18)$$

The first form shows specifically the three components of $\mathbf{e}$ in the directions of $\mathbf{r}, \mathbf{u}, \mathbf{a}$, while the second conveniently separates the acceleration-dependent part of $\mathbf{e}$ from the rest. We note, above all, that the entire acceleration-dependent part of $\mathbf{e}$ is $O(\frac{1}{r})$ while the rest is $O(\frac{1}{r^2})$. It is also of interest to note that the acceleration-*independent* part of $\mathbf{e}$ points away from where the charge *would* be at the instant of observation had it proceeded with uniform velocity from the retarded event [cf. after (41.3)].

We could now proceed to calculate $\mathbf{b}$ similarly, but it is more convenient to establish at once what we would in any case find after that calculation, namely that

$$c\mathbf{b} = \frac{1}{r}(\mathbf{r} \times \mathbf{e}). \quad (40.19)$$

The proof of this goes as follows—where we temporarily write $p = Qc^2/D^3$ to reduce ballast, and also use the last result of Exercise A(8), and (39.3)–(39.5):

$$\begin{aligned}
(\mathbf{r} \times \mathbf{e})^h &= e^{hij}r_i e_j = -e^{hij}R_i E_{0j} \\
&= -e^{0hij}R_i p(R_0 \tilde{U}_j - R_j \tilde{U}_0) \\
&= r\varepsilon^{0hij} p R_{[i}\tilde{U}_{j]} = \tfrac{1}{2}r\varepsilon^{0hij} E_{ij} \\
&= rB^{0h} = rcb^h,
\end{aligned}$$

which is equivalent to (40.19).

If we look specifically at the acceleration-dependent part of the field, say $\tilde{\mathbf{e}}$ and $\tilde{\mathbf{b}}$, we see from (40.18) that $\tilde{\mathbf{e}} \perp \mathbf{r}$, while from (40.19), $c\tilde{\mathbf{b}} \perp \tilde{\mathbf{e}}$, $\mathbf{r}$ and $c\tilde{b} = \tilde{e}$. So the acceleration field is a *null* field (i.e. satisfying $X = Y = 0$, cf. Section 39), and orthogonal to $\mathbf{r}$.

41. Field of a uniformly moving charge

When a charge q moves uniformly through an inertial frame S, it makes good sense to look at its entire field at *one instant* in S, which might as well be $t = 0$. Suppose the charge moves with velocity $\mathbf{u}$ along the x-axis and passes the origin O at $t = 0$. Let R be its retarded position relative to the general point P at which we shall calculate the field. Also write $\mathbf{d} = \overrightarrow{RO}$, $\mathbf{r} = \overrightarrow{RP}$, $\mathbf{r}_0 = \overrightarrow{OP}$, and let ϕ and θ be the

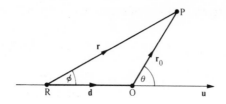

F$_\text{IG}$.   18

angles between the x-axis and $\mathbf{r}$ and $\mathbf{r}_0$, respectively (see Fig. 18).

Now from (40.18) we have, when $\mathbf{a} = 0$,

$$\mathbf{e} = \frac{Qc^3}{Z^3\gamma^2}\left(\mathbf{r} - \frac{r}{c}\mathbf{u}\right), \quad Z = cr - \mathbf{r}\cdot\mathbf{u} \tag{41.1}$$

where for convenience we have introduced the symbol Z. Since the charge travels at speed u from R to O in the time that light would travel from R to P, we have

$$\frac{d}{u} = \frac{r}{c}, \tag{41.2}$$

and so

$$\mathbf{r} - \frac{r}{c}\mathbf{u} = \mathbf{r} - \mathbf{d} = \mathbf{r}_0. \tag{41.3}$$

In conjunction with (41.1), this leads to the surprising result that the field at P at $t = 0$ points away from where the charge is *at $t = 0$*, and not from where it was when it presumably *caused* that field.

Next consider the simple identity

$$(cr - r\mathbf{u})^2 - (cr - \mathbf{r}\cdot\mathbf{u})^2 \equiv (ru\sin\phi)^2, \tag{41.4}$$

whose validity can be checked at once. Since, by elementary geometry, $r\sin\phi = r_0\sin\theta$, (41.4) tells us that

$$(cr_0)^2 - Z^2 = u^2r_0^2\sin^2\theta,$$

and so, with (41.3), (41.1) finally becomes

$$\mathbf{e} = \frac{Q\mathbf{r}_0}{\gamma^2 r_0^3[1 - (u^2/c^2)\sin^2\theta]^{3/2}}, \quad Q = \frac{q}{4\pi\varepsilon_0}. \tag{41.5}$$

Though the result (40.19) is of course still true, it is well here to eliminate its dependence on the retarded position as follows:

$$c\mathbf{b} = \frac{1}{r}(\mathbf{r} \times \mathbf{e}) = \frac{1}{r}(\mathbf{d} + \mathbf{r}_0) \times \mathbf{e} = \frac{1}{r}(\mathbf{d} \times \mathbf{e})$$

$$= \frac{1}{c}(\mathbf{u} \times \mathbf{e}). \tag{41.6}$$

This completes our derivation of the field of a uniformly moving charge. However, had that been our *only* purpose, we could have found a quicker way: see, for example, Exercise VI(10).

Note that both the electric and magnetic field strengths in any fixed direction from the charge fall off as $1/r_0^2$. Also of interest is the angular dependence of the strength of $\mathbf{e}$; it is strongest in a plane at right angles to $\mathbf{u}$, and weakest fore and aft.

We can construct a remarkable model of a uniformly moving charge in which the density of the lines of $\mathbf{e}$ represents field strength as usual. Consider a point charge at the origin of its rest frame S′, and imagine electrically inert 'spikes' attached to it completely isotropically to represent the $\mathbf{e}$-lines emanating from the charge so that, in the usual way, their density equals the field strength at any point. Next let us look at this spiky model in the general frame S through which it travels, say, with speed u. It suffers the usual Lorentz contraction and the spikes bunch up at the sides. We shall now prove that in S they still exactly represent the field by their direction *and* density! In S′ the solid angle of a thin pencil of lines, making an angle θ' with the x'-axis and having x-cross-sectional area dA at (x', y', z'), is given by

$$d\Omega' = dA \cos\theta'/r'^2 = dAx'/r'^3,$$

with $r'^2 = x'^2 + y'^2 + z'^2$ (see Fig. 19(a)). In S the corresponding solid

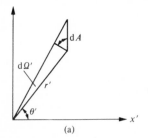

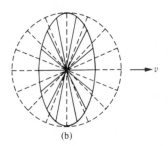

(a)   (b)

Fɪɢ.   19

angle is given by $d\Omega = dAx/r^3$, since dA has the same measure in both frames. But, by length contraction,

$$r'^2 = \gamma^2 x^2 + y^2 + z^2 = \gamma^2 r^2 - (\gamma^2 - 1)(y^2 + z^2)$$
$$= \gamma^2 r^2 [1 - (u^2/c^2)\sin^2\theta], \tag{41.7}$$

and so

$$\frac{d\Omega'}{d\Omega} = \frac{x'r^3}{xr'^3} = \frac{\gamma r^3}{r'^3} = \frac{1}{\gamma^2 [1 - (u^2/c^2)\sin^2\theta]^{3/2}}. \tag{41.8}$$

Comparing this with (41.5) we see that

$$e = \frac{Qd\Omega'}{r^2 d\Omega} = \frac{n}{d\Sigma}, \tag{41.9}$$

where $Qd\Omega' = n$ is the number of field lines conventionally assigned to $d\Omega'$ in S' and $r^2 d\Omega = d\Sigma$ is the normal cross-sectional area of that same bundle of lines in S. So we see that in S too the field strength can be read off as the density of the field lines represented by our spikes. In other words, the e-field lines transform like rigid spikes attached to the charge (see Fig. 19(b)). For example, the lateral field of a very fast-moving charge is almost a shock wave: it is not only very large, but also essentially 'radiative' ($cb = e$, $\mathbf{b} \perp \mathbf{e}$).

Of course, the result illustrated by Fig. 19(b) is purely a consequence of the laws of electrodynamics and can be obtained without the explicit use of relativity. Lorentz so obtained it, and thereon based an 'explanation' of the length contraction of material bodies: if the electromagnetic fields of the fundamental charges 'contract', he argued, then so must all matter, if it is made up of such charges. (Lorentz's argument, perforce, ignored the existence of the nuclear forces, uncharged fundamental particles, etc.)

42. The electromagnetic energy tensor

In this section we show how the electromagnetic field itself can be credited with possessing energy, momentum, and stress, in order to permit extensions of the various conservation laws of matter to situations where matter interacts with electromagnetic fields. Consider, for example, two identical point-masses carrying identical charges and being released from rest simultaneously on the x-axis of an inertial frame S. They will move apart symmetrically, and so momentum is conserved. Mechanical energy is *not* conserved, as both particles speed up. This is usually 'explained' by saying that while the particles gain kinetic energy, they lose *potential* energy. But now ride past this system in the usual second frame S'. Originally both particles have equal constant velocities in S'. Then suddenly the velocity of the 'right' particle decreases, while that of the other is still unchanged (relativity of simultaneity!). What about momentum conservation here? Is there such a thing as 'potential momentum'? Perhaps one could introduce such a concept, in analogy with potential energy, which is really nothing but a useful 'book-keeping' device. But physically it is more satisfactory to credit the field *itself* with whatever momentum or energy is required to 'balance the books'. It turns out that there is a very elegant way of doing this. Moreover, it is not really a matter of *choice* whether to place the energy, momentum, etc., here or there. If we accept Einstein's general relativity, only one location can be 'correct', for that in turn causes local spacetime curvature which, at least in principle, is directly observable.

So consider an electromagnetic field $E_{\mu\nu}$ and in it a distribution of charged 'fluid' subject to no other external forces. This fluid constitutes a convection current J^μ, say. In order to find the Lorentz four-force $\tilde{K}_\mu$ acting on a unit *proper* volume of fluid we multiply (38.1) by n_0, the number of elementary charges q per unit proper volume; then, since $n_0 q \mathbf{U} = \rho_0 \mathbf{U} = \mathbf{J}$ [cf. (38.17)], this yields

$$\tilde{K}_\mu = E_{\mu\nu} J^\nu / c. \tag{42.1}$$

Certain important results can be obtained by transforming the right-hand side of this equation into a divergence by use of Maxwell's equations. Substituting for J^ν from (38.3) gives

$$\tilde{K}_\mu = \varepsilon_0 E_{\mu\nu} E^{\sigma\nu}{}_{,\sigma} = \varepsilon_0 \left[(E_{\mu\nu} E^{\sigma\nu})_{,\sigma} - E_{\mu\nu,\sigma} E^{\sigma\nu} \right]. \tag{42.2}$$

For the last term we find, using first the antisymmetry of $E^{\sigma\nu}$, then the field equation (38.7), and lastly the see-saw rule,

$$E_{\mu\nu,\sigma} E^{\sigma\nu} = \tfrac{1}{2} (E_{\mu\nu,\sigma} - E_{\mu\sigma,\nu}) E^{\sigma\nu}$$
$$= \tfrac{1}{2} E_{\nu\sigma,\mu} E^{\nu\sigma} = \tfrac{1}{4} (E_{\nu\sigma} E^{\nu\sigma})_{,\mu}.$$

When we substitute this into (42.2) (and change a few dummy pairs), we get the desired expression

$$\tilde{K}_\mu = \varepsilon_0 \left[E_{\mu\sigma} E^{\nu\sigma} - \tfrac{1}{4} \delta_\mu^\nu (E_{\rho\sigma} E^{\rho\sigma}) \right]_{,\nu}. \tag{42.3}$$

This can be written as

$$\tilde{K}_\mu = - M^\nu_{\mu,\nu}, \tag{42.4}$$

thereby defining the tensor

$$M^{\nu}_{\mu} = -\varepsilon_0 [E_{\mu\sigma} E^{\nu\sigma} - \tfrac{1}{4}\delta^{\nu}_{\mu}(E_{\rho\sigma} E^{\rho\sigma})], \tag{42.5}$$

called the *energy tensor* of the electromagnetic field for reasons that will become apparent presently. From the result of Exercise A (20) we immediately obtain the following alternative expression for M^{ν}_{μ}:

$$M^{\nu}_{\mu} = -\tfrac{1}{2}\varepsilon_0 [E_{\mu\sigma} E^{\nu\sigma} + B_{\mu\sigma} B^{\nu\sigma}], \tag{42.6}$$

which demonstrates its symmetry in **e** and c**b**. In fact, M^{ν}_{μ} is also symmetric in the usual sense,

$$M^{\mu\nu} = M^{\nu\mu}, \tag{42.7}$$

for $E^{\mu}_{\ \sigma} E^{\nu\sigma} = E^{\mu\sigma} E^{\nu}_{\ \sigma} = E^{\nu}_{\ \sigma} E^{\mu\sigma}$ and similarly for the dual term.

Before proceeding further we shall work out the components of M^{ν}_{μ} in the general frame. If we define the symbols σ, g_i, p_{ij} by either of the two equivalent schemes

$$M^{\nu}_{\mu} = \begin{pmatrix} \sigma & -cg_1 & -cg_2 & -cg_3 \\ cg_1 & -p_{11} & -p_{12} & -p_{13} \\ cg_2 & -p_{21} & -p_{22} & -p_{23} \\ cg_3 & -p_{31} & -p_{32} & -p_{33} \end{pmatrix},$$

$$M^{\mu\nu} = \begin{pmatrix} \sigma & cg_1 & cg_2 & cg_3 \\ cg_1 & p_{11} & p_{12} & p_{13} \\ cg_2 & p_{21} & p_{22} & p_{23} \\ cg_3 & p_{31} & p_{32} & p_{33} \end{pmatrix}, \tag{42.8}$$

and write δ_{ij} for the three-dimensional Kronecker delta, we easily find from (42.6) the following expressions:

$$\sigma = \tfrac{1}{2}\varepsilon_0 (e^2 + c^2 b^2), \tag{42.9}$$

$$\mathbf{g} = (g_1, g_2, g_3) = \varepsilon_0 (\mathbf{e} \times \mathbf{b}), \tag{42.10}$$

$$p_{ij} = p_{ji} = -\varepsilon_0 [e_i e_j + c^2 b_i b_j - \tfrac{1}{2}\delta_{ij}(e^2 + c^2 b^2)]. \tag{42.11}$$

Now if the fluid moves with velocity **u**, and if $\tilde{\mathbf{k}}$ is the Lorentz three-force on it per unit proper volume and **k** that per unit volume, we have $\mathbf{k} = \gamma(u)\tilde{\mathbf{k}}$, because the moving volume is contracted by a γ-factor. Also since the Lorentz force is pure, we have [cf. (35.12)] for the four-force per unit proper volume

$$\tilde{K}_{\mu} = \gamma(u)(\tilde{\mathbf{k}} \cdot \mathbf{u}/c, -\tilde{\mathbf{k}}) = (\mathbf{k} \cdot \mathbf{u}/c, -\mathbf{k}). \tag{42.12}$$

Then if we set $\mu = 0$ in (42.4), we get

$$\mathbf{k} \cdot \mathbf{u} = -\frac{\partial \sigma}{\partial t} - \operatorname{div} c^2 \mathbf{g}. \tag{42.13}$$

But we know from (35.8) (with $m_0 = $ constant) that $\mathbf{k} \cdot \mathbf{u}$ is the rate at which the field does work on a unit volume of fluid. It is therefore natural to regard σ as the *energy density* of the field, and $c^2 \mathbf{g}$ as its *energy current density*, so that (42.13) becomes an *equation of continuity*, i.e. of (energy) conservation. For it then states that the expenditure of field energy $(\mathbf{k} \cdot \mathbf{u})$ per unit volume is 'fully paid for' by the decrease of internal energy $(-\partial \sigma/\partial t)$ plus the influx of external energy $(-\operatorname{div} c^2 \mathbf{g})$. The energy current density is also known as the *Poynting vector*.

To motivate our next step, consider some fluid having energy density σ and velocity $\mathbf{u}$. Then the energy current density $\boldsymbol{\varepsilon}$ will be given by $\boldsymbol{\varepsilon} = \sigma \mathbf{u}$ [cf. (38.18) for the electric analogue]. On the other hand, because of $E = mc^2$, the momentum density $\mathbf{g}$ of the fluid will be given by $\mathbf{g} = (\sigma/c^2)\mathbf{u}$, so that

$$\boldsymbol{\varepsilon} = c^2 \mathbf{g}. \tag{42.14}$$

This relation holds even if there are several independent energy currents flowing with different velocities, for it holds for each one and consequently also for the resultant. By analogy, having recognized $c^2 \mathbf{g}$ as the energy current density of the *field*, it follows that we must regard $\mathbf{g}$ as its *momentum density*.

If we next set $\mu = i$ in (42.4), we find

$$-k_i = \frac{\partial g_i}{\partial t} + \frac{\partial p_{ij}}{\partial x^j}. \tag{42.15}$$

Since $\mathbf{k}$ is the force of the field on the matter, $-\mathbf{k}$ can be regarded as the force of the matter on the field. So $-\mathbf{k}$ should be the rate at which *field* momentum is generated inside a unit volume. The first term on the right represents the increase of the internal momentum. The divergence-like second term should therefore represent the *outflux* of field momentum through the surface. Maxwell interpreted this term as due to the field's *internal stresses*. (Recall: force = rate of absorption of momentum.) Specifically, according to Maxwell, p_{ij} is the i-component of the total force which the *field* (!) on the negative side of a unit area normal to the x^j-axis exerts upon the *field* on the positive side. For consider a small cube of the field with its edges, of length dl, parallel to the coordinate axes. The force on the face open to the negative x^j-axis is then $p_{ij}dl^2$, and—by the equality of action and reaction—that on the opposite side will be

$$- (p_{ij} + \frac{\partial p_{ij}}{\partial x^j} dl) dl^2 \quad \text{(no sum)}.$$

The net force (and thus the momentum influx) from this pair of faces is therefore

$$- \frac{\partial p_{ij}}{\partial x^j} dl^3 \quad \text{(no sum)}.$$

So if we sum the contributions from all three pairs of faces and divide by dl^3, we find that the net influx of field momentum per unit volume is

$$- \frac{\partial p_{ij}}{\partial x^j}. \tag{42.16}$$

With that, (42.15) becomes the *equation of continuity* for the field momentum. Thus we see that in order to achieve overall conservation of energy and momentum where matter and fields interact, we must credit the field itself not only with energy and momentum of its own, but also with stress. In fact, the simple law of energy conservation (42.13) in *all* inertial frames *implies* momentum conservation with its involvement of stress, according to (42.15): we cannot have the one without the other, as follows from the zero component lemma as applied to the vector $K_\mu + M^\nu_{\mu,\nu}$. The situation in this respect is quite analogous to that obtaining in particle mechanics [cf. end of Section 27]. The realization that the energy, momentum, and Maxwell stress of the field combine to form a single 'energy tensor' was one of the great achievements of Minkowski. (Hence our notation $M^{\mu\nu}$: for Minkowski *and* Maxwell.)

We may note that implicit in our work above is a first instance of Einstein's mass–energy equivalence outside of mechanics. It consists in the fact that **g** (from the first row of $M^{\mu\nu}$) acts as momentum density while c^2 **g** (from the first column) acts as energy current density, in conformity with our relation (42.14) which crucially depends on $E = mc^2$. We also note how the symmetry of $M^{\mu\nu}$ (first row equals first column) is connected with this. (The symmetry $p_{ij} = p_{ji}$ is analogous to the well-known symmetry of the stress tensor in the classical theory of elasticity.)

Finally we observe that p_{ij} can be used for the practical purpose of calculating the total electromagnetic force on a region of field, e.g. that occupied by a material body: one need merely evaluate the surface integral of p_{ij} over the boundary. The p_{ij} also imply certain 'elastic' properties of the field lines. As a simple example, consider a pure e-field which is parallel to the x-axis at the point of interest. Then, from (42.11), $p_{11} = -\frac{1}{2}\varepsilon_0 e^2$. But p_{11} is pressure in the x-direction. This leads to the concept that there is *tension* along the electric field lines. Such tension 'explains', for example, the attraction between unequal charges. Similarly, $p_{22} = +\frac{1}{2}\varepsilon_0 e^2$. So there is positive pressure at right angles to the electric field lines, tending to separate them. This 'explains', for example, the repulsion between equal charges.

43. Electromagnetic waves

We have already seen in (38.14) that the electromagnetic field $E_{\mu\nu}$ in vaccum satisfies the wave equation. We have also seen—in fact, in getting that result—that wave solutions of the potential Φ^μ automatically give rise to wave solutions of the field. Let us therefore begin by looking at *plane, sinusoidal* ('monochromatic') waves of the potential, which we shall then translate into waves of the field. By Fourier analysis, *sums* of such waves form the general plane wave.

By reference to (24.12), (24.13), and (24.6), we can write, for a plane wavetrain of speed c and frequency v in the direction $\mathbf{n}$,

$$\Phi^\mu = \operatorname{Re} \Psi^\mu, \quad \Psi^\mu = D^\mu \exp(iM_\sigma x^\sigma), \tag{43.1}$$

where

$$M_\sigma = (m, -\mathbf{m}) = (2\pi/c)N_\sigma = (2\pi/c)v(1, -\mathbf{n}), \tag{43.2}$$

and where $D^\mu = (D^0, \mathbf{d})$ is a constant complex vector called the *polarization vector*. The potential (43.1) automatically satisfies the wave equation $\Box\Phi^\mu = \operatorname{Re}\Box\Psi^\mu = 0$, by the second of the following equations and by the nullity of M_μ:

$$\Psi_{\mu,\nu} = iD_\mu M_\nu \exp(\ldots), \ \Psi_{\mu,\nu\rho} = -D_\mu M_\nu M_\rho \exp(\ldots). \tag{43.3}$$

From the first of these equations we see that the gauge condition $\Phi^\mu_{,\mu} = \operatorname{Re}\Psi^\mu_{,\mu} = 0$—without which a solution of the wave equation is not a Maxwell potential—here reduces to

$$M_\mu D^\mu = 0. \tag{43.4}$$

Next we obtain the field $E_{\mu\nu}$ from Φ_μ by use of (43.3) (i):

$$E_{\mu\nu} = \Phi_{\nu,\mu} - \Phi_{\mu,\nu} = \text{Re}[i(M_\mu D_\nu - M_\nu D_\mu)\exp(...)]. \quad (43.5)$$

Now this as well as the gauge condition (43.4) is unaffected by a change $D^\mu \to D^\mu + \text{constant} \times M^\mu$ (the latter because of the nullity of M^μ).[1] Since $M^0 \neq 0$ we can use this freedom to make $D^0 = 0$, whereupon (43.4) reads

$$\mathbf{n} \cdot \mathbf{d} = 0. \quad (43.6)$$

Then (43.5) yields the following expressions for $\mathbf{e} = E_{0i}$ and $c\mathbf{b}$ $= (E_{32}, E_{13}, E_{21})$:

$$\mathbf{e} = -\text{Re}[im\mathbf{d}\exp(iM_\sigma x^\sigma)] \quad (43.7)$$

$$c\mathbf{b} = -\text{Re}[im \times \mathbf{d}\exp(iM_\sigma x^\sigma)] = \mathbf{n} \times \mathbf{e}. \quad (43.8)$$

From (43.6) and (43.7) we see that $\mathbf{e}$ is orthogonal to $\mathbf{n}$, and then (43.8) shows that $cb = e$ and that $\mathbf{e}, \mathbf{b}, \mathbf{n}$ form a right-handed orthogonal triad, in that order. It is also clear that for the sum of any number of waves of this type the same relations are true: $\mathbf{n} \cdot \Sigma\mathbf{e} = \Sigma\mathbf{n} \cdot \mathbf{e} = 0$ and $\Sigma c\mathbf{b} = \Sigma(\mathbf{n} \times \mathbf{e}) = \mathbf{n} \times \Sigma\mathbf{e}$.

It is not hard to show from (43.7) that at a fixed point in space the $\mathbf{e}$ vector rotates with its tip in general following an ellipse. Similarly, along any ray at one instant, the tip of the $\mathbf{e}$ vector also traces out an ellipse, but in the opposite sense. In the special case when $\mathbf{d}$ is real we have, from (43.7),

$$\mathbf{e} = m\mathbf{d}\sin(...). \quad (43.9)$$

Here the tip of $\mathbf{e}$ performs simple harmonic motion in a fixed direction at all points of space. One calls such a wave *linearly polarized*. (Much the same happens when $\mathbf{d}$ is purely imaginary.) Evidently by a suitable superposition of two such waves one can produce a *circularly polarized* wave, where the tip of $\mathbf{e}$ follows a circle. This occurs in (43.7) when $\mathbf{d} = \mathbf{d}_1 + i\mathbf{d}_2$ with $\mathbf{d}_1$ and $\mathbf{d}_2$ real, orthogonal, and equal in magnitude.

Now choose the z-axis parallel to $\mathbf{n}$, so that, at any event in the wave,

$$\mathbf{e} = e(\cos\alpha, \sin\alpha, 0), \quad c\mathbf{b} = e(-\sin\alpha, \cos\alpha, 0) \quad (43.10)$$

for some angle α. From this and (42.8)–(42.11) it is now easy to see that the following are the only non-zero components of the energy tensor at the given event:

$$M^{00}(=\sigma) = M^{03} = M^{30}(=cg) = M^{33}(=p) = \varepsilon_0 e^2, \quad (43.11)$$

where we have written p for p_{33}, the longitudinal pressure of the wave. Since in these coordinates

$$N^\mu = v(1, 0, 0, 1) \tag{43.12}$$

[cf. (43.2)], we can condense (43.11) into the equation

$$M^{\mu\nu} = \frac{\varepsilon_0 e^2}{v^2} N^\mu N^\nu. \tag{43.13}$$

This must be true in all inertial frames. But then it follows from a variation of the quotient rule that e/v is a scalar, so the values of e and v in two arbitrary inertial frames S and S' are related by

$$\frac{e'}{e} = \frac{v'}{v}. \tag{43.14}$$

This in conjunction with (43.11) leads to the following more extensive set of proportionalities:

$$\frac{\sigma'}{\sigma} = \frac{g'}{g} = \frac{p'}{p} = \frac{e'^2}{e^2} = \frac{v'^2}{v^2}. \tag{43.15}$$

Since the ratio v'/v is clearly constant over space and time, it follows that the *averages* over space *or* time of the various quantities appearing in (43.15) obey the same relations as the instantaneous values:

$$\frac{\langle \sigma' \rangle}{\langle \sigma \rangle} = \frac{\langle g' \rangle}{\langle g \rangle} = \frac{\langle p' \rangle}{\langle p \rangle} = \frac{\langle e'^2 \rangle}{\langle e^2 \rangle} = \frac{v'^2}{v^2}. \tag{43.16}$$

A problem already considered by Einstein in his 1905 paper on special relativity was the transformation of the total energy in a 'wave complex'. Let us therefore look at a given portion of a wavetrain, say that between two successive wave crests a wavelength λ apart, and laterally bounded by a tube of rays. Since the lateral dimensions of this volume are the same in all inertial frames—by our results on supersnapshots at the end of Section 18—the volume is proportional to λ and thus inversely proportional to v. So the total energy E of this complex is proportional to $\langle \sigma \rangle/v$. But we have just seen in (43.16) that $\langle \sigma \rangle/v^2$ is an invariant. Hence it follows—and was considered 'remarkable' by Einstein—that E is directly proportional to v. No doubt this played a role in the development of his idea of the photon.

Another elementary problem of interest is the radiation pressure on a blackened (perfectly absorbing) plate in the path of the waves, say at right angles to the rays. We can argue by momentum conservation: the force on the plate equals the rate at which momentum is absorbed by it. Since a volume c of radiation is absorbed in unit time by a unit area of plate, an amount $c \langle g \rangle$ of momentum is absorbed, and so, by (43.11), the average pressure is $\varepsilon_0 \langle e^2 \rangle$. If, instead, the plate is a perfect reflector, the momentum $c \langle g \rangle$ is changed into $-c \langle g \rangle$ so that $2c \langle g \rangle$ is absorbed and the pressure is twice as big as before. Note that in the first case the radiation pressure is also equal to the 'internal' pressure $\langle p \rangle$ of the wave, but this is not always so. (For example, if a plate moves though a 'photon gas' relative to its 'rest frame', it will clearly suffer different pressures on its two sides, whereas there is only one 'field' pressure.)

We may finally remark that all our results above on *plane* monochromatic waves might well be expected to hold locally also for *arbitrary* monochromatic waves, as long as the 'geometrical optics' limit of high frequency is applicable [cf. after (24.2)]. For then any small wave region will contain sufficiently many complete waves to approximate to a plane wavetrain.

[1] This artifice is taken from N. Woodhouse, mimeographed *Notes on Special Relativity*, Oxford, 1980, Section 6.8.

Exercises VI

1. (i) A particle of rest mass m and charge q is injected at velocity **u** into a constant pure magnetic field **b** at right angles to the field lines. Use the Lorentz force law (38.16) to establish that the particle will trace out a circle of radius $mu\gamma(u)/qb$ with period $2\pi m\gamma(u)/qb$. [It was the γ-factor in the period that necessitated the development of synchrotrons from cyclotrons, at whose energies the γ was still negligible.]

(ii) If the particle is injected into the field with the same velocity but at an angle $\theta \neq \pi/2$ to the field lines, prove that the path is a helix, of smaller radius, but that the period for one complete cycle is the same as before.

2. Prove that the validity of the first Maxwell equation, (38.20)(i), in *all* inertial frames implies that of the second, (38.20) (ii). Note how this provides justification for Maxwell's 'displacement current' $\partial \mathbf{e}/\partial t$. [After having covered equation (39.11) in the text, you can similarly show that the validity of (38.21) (i) in all inertial frames implies that of (38.21) (ii).]

3. Mainly for future reference, write down the inverse transformation equations to (39.1) and (39.2).

4. Prove, by any method, that the electric field $\mathbf{e}$ at a point P due to an infinite straight line distribution of static charge, λ per unit length, is given by $\mathbf{e} = \lambda\mathbf{r}/2\pi\varepsilon_0 r^2$, where $\mathbf{r}$ is the perpendicular vector-distance of P from the line. Deduce, by transforming to a frame in which this line moves, that the magnetic field $\mathbf{b}$ at P due to an infinitely long straight current $\mathbf{i}$ is given by $\mathbf{b} = (\mathbf{i} \times \mathbf{r})/2\pi\varepsilon_0 c^2 r^2$. Check this with (38.20) (ii).

5. Verify that the complex three-vector $\mathbf{k} = \mathbf{e} + ic\mathbf{b}$ transforms under the transformation (39.1), (39.2) of $\mathbf{e}$ and $\mathbf{b}$ as does a three-vector under a rotation about the x-axis through an imaginary angle. [*Hint*: Exercise I(12).] Deduce that $c^2b^2 - e^2$ and $\mathbf{e} \cdot \mathbf{b}$ are the *only* invariants of the electromagnetic field.

6. If $(\mathbf{e}, \mathbf{b})$ and $(\mathbf{e}', \mathbf{b}')$ are two different electromagnetic fields, prove that $c^2\mathbf{b} \cdot \mathbf{b}' - \mathbf{e} \cdot \mathbf{e}'$ and $\mathbf{e} \cdot \mathbf{b}' + \mathbf{b} \cdot \mathbf{e}'$ are invariants.

7. Prove the 'zero component' lemma for an antisymmetric tensor $T^{\mu\nu}$: if any *one* of its off-diagonal components is zero in all inertial frames, then the entire tensor is zero.

8. In a frame S there is a uniform electric field $\mathbf{e} = (0, a, 0)$ and a uniform magnetic field $c\mathbf{b} = (0, 0, 5a/3)$. A particle of rest mass m and charge q is released from rest on the x-axis. What time elapses before it returns to the x-axis? [*Answer*: $74\pi cm/32aq$. *Hint*: look at the situation in a frame in which the electric field vanishes.]

9. Obtain the Liénard–Wiechert potentials (40.8) of an arbitrarily moving charge q by the following alternative method: Assume, first, that the charge moves uniformly and that in its rest frame the potential is given by (40.2). Then transform this to the general frame, using the four-vector property of Φ^μ. Finally extend the result to an arbitrarily moving charge by the argument we used after (40.2). [*Hint*: if the separation $(ct, \mathbf{r})$ between two events satisfies $r = ct$ in one frame, it does so in all frames.]

10. Obtain the field (41.5), (41.6) of a uniformly moving charge $q[= (4\pi\varepsilon_0)^{-1}Q]$ by the following alternative method: Assume that the field in the rest frame S′ of the charge is given by

$$\mathbf{e}' = (Q/r'^3)(x', y', z'), \quad \mathbf{b}' = 0, \quad r'^2 = x'^2 + y'^2 + z'^2,$$

then transform this field to the usual second frame S at $t = 0$. [*Hint*: obtain $\mathbf{b} = \mathbf{u} \times \mathbf{e}/c^2$ from (39.2); from the inverse of (39.1) obtain $\mathbf{e} = (Q\gamma/r'^3)(x, y, z)$; finally use (41.7).]

11. In a frame S, two identical particles with electric charge q move abreast along lines parallel to the x-axis, a distance r apart and with velocity v. Determine the force, in S, that each exerts on the other, and do this in two ways: (i) by use of (41.5), (41.6), and (38.16); and (ii) by transforming the Coulomb force from the rest frame S' to S by use of the four-vector property of (35.5), with $dm'/dt' = 0$ (why?). Note that the force is smaller than in the rest frame, while each mass is greater. Here we see the dynamical reasons for the 'relativistic focusing' effect whose existence we predicted by purely kinematic considerations in Section 11. Do these dynamical arguments lead to the exact expected time dilation of an 'electron cloud clock'? Also note from (i) that as $v \to c$ the electric and magnetic forces each become infinite, but that their effects cancel.

12. Instead of the equal charges moving abreast as in the preceding exercise, consider now two *opposite* charges moving along parallel lines at the same constant velocity but not abreast. By both suggested methods determine the forces acting on these charges, and show that they do not act along the line joining the charges (e.g. a non-conducting rod) but, instead, constitute a couple tending to turn that join into orthogonality with the line of motion. [Trouton and Noble, in a famous experiment in 1903, unsuccessfully tried to detect this couple on charges at rest in the laboratory, which they presumed to be flying through the ether. The fact that the rod's reaction could also be not in line with it was unsuspected and the result seemed puzzling. However, it contributed to the later acceptance of relativity.]

13. If $e \cdot b \neq 0$, prove that there are infinitely many frames (with common relative direction of motion, and only those) in which e is parallel to b; precisely one of these moves in the direction $e \times b$, its velocity being given by the smaller root of the quadratic $\beta^2 - R\beta + 1 = 0$, where $\beta = v/c$ and $R = (e^2 + c^2 b^2)/|e \times c b|$. For the reality of β it is necessary to show that $R > 2$. [*Hint*: use (39.1), (39.2), and choose the spatial axes judiciously.]

From the interpretation of σ and $c^2 g$ in Section 42 it might have been expected, wrongly, that the required value of β is $2/R$. But the fact that there are infinitely many frames in which the energy current of the field vanishes (except in a null field) indicates that we cannot generally ascribe a unique velocity to this current.

14. Prove that the electromagnetic energy tensor satisfies the following two identities:

$$M^\mu_\mu = 0, \quad M^\mu_\sigma M^\sigma_\nu = (I\varepsilon_0/2)^2 \delta^\mu_\nu,$$

where $I^2 = (c^2b^2 - e^2)^2 + 4c^2(\mathbf{e}\cdot\mathbf{b})^2$. [*Hint:* It may be easiest to establish the second identity in a particular frame, e.g. if $I \neq 0$, in one in which $\mathbf{e}$ is parallel to $\mathbf{b}$. (We regard the vanishing of $\mathbf{e}$ or $\mathbf{b}$ as a particular case of this.) For the case $I = 0$, appeal to continuity.]

15. Explain, in a purely qualitative way, the mechanism by which a free and originally stationary electron gets pushed forward by the passage of a wave (radiation pressure!). In order to avoid a paradoxical *backward* push, show that we must have $\mathbf{b} = \mathbf{n} \times \mathbf{e}$ rather than $\mathbf{b} = -\mathbf{n} \times \mathbf{e}$. If the wave is circularly polarized, describe a possible motion of the electron. Deduce also that circularly polarized light carries angular momentum (in fact, of amount $\sigma\omega$ per unit volume, if ω is the rate of turning of the field).

16. The apparent brightness b of a distant luminous source (which can be idealized as a point source) is the rate at which radiant energy from it is received on a unit area perpendicular to the line of sight; the absolute brightness B is defined as the apparent brightness at unit distance from the source in its rest frame. Prove that for two momentarily coincident observers the apparent brightnesses of the same source are in the ratio of the squares of the observed frequencies: $b'/b = v'^2/v^2$. Hence or otherwise prove that $b = (B/R^2)(v^4/v_0^4)$, where R is the distance of the source from the oberserver in the observer's inertial frame at the time the light he measures was emitted, and v_0 is its proper frequency. [*Hint:* for the two frames of Fig. 10(*b*), $R \sin \alpha = R' \sin \alpha'$.]

17. Give reasons why in a disordered (i.e. random) distribution of pure radiation (a 'photon gas') the electromagnetic field components will satisfy the following relations on the (time) average:

(i) $e_1^2 = e_2^2 = e_3^2$, $\quad b_1^2 = b_2^2 = b_3^2$,

(ii) $e_1e_2 = e_2e_3 = e_3e_1 = 0$, $\quad b_1b_2 = b_2b_3 = b_3b_1 = 0$,

(iii) $e_2b_3 - e_3b_2 = e_3b_1 - e_1b_3 = e_1b_2 - e_2b_1 = 0$.

Deduce that the only non-zero components of the averaged energy tensor can then be written as $M^{00} = \sigma_0$, $M^{11} = M^{22} = M^{33} = p$, where $3p = \sigma_0$.

18. In a frame S there is a random distribution of radiation with energy tensor as in the preceding exercise. Prove that the radiation pressures p_f and p_b on the front and back, respectively, of a balckened plate fixed in the usual second frame S' at right angles to the x'-axis,

are given by

$$p_f = \tfrac{1}{2} p \frac{(1+\beta)^2}{1-\beta}, \quad p_b = \tfrac{1}{2} p \frac{(1-\beta)^2}{1+\beta},$$

where $\beta = v/c$. Verify that $p_f + p_b = M^{1'1'}$. [*Hint*: Use the result of Exercise III(12). The contribution to the pressure at any point of the plate must be summed over all incident angles, and it must be remembered that equal solid angles in S and not in S′ contain 'equal numbers of photon tracks' to the point in question.]

VII

RELATIVISTIC MECHANICS OF CONTINUA

44. Introduction

We would expect any continuous distribution of a matter—a gas, a liquid, a solid—here generically called a *fluid*, to possess an energy density, a momentum density, and internal stresses. Since for the electromagnetic field, as we have seen, these quantities combine to form a four-tensor, which must be a consequence of their basic transformation properties, we would surely expect the same for a fluid. And indeed this is true: relativistic fluid dynamics is characterized by an energy tensor similar to that of the electromagnetic field which also satisfies a 'divergence equation' similar to that of the electromagnetic field. But, it may well be asked, why bother with relativistic fluid dynamics, when surely we cannot accelerate macroscopic fluids to the point at which relativistic effects become important? (For example, one never bothers to work out the relativistic Doppler effect for *sound* waves, though in principle that also exists.) There are at least three good answers to this question. First, we are driven to it for purely logical reasons: fluid mechanics is logically more basic than particle mechanics, inasmuch as the specialization from the former to the latter is much more direct than the opposite procedure. Secondly, it turns out that relativistic fluid dynamics is the basic ingredient of general relativity, where the energy tensor of the fluid serves as the source of the gravitational field—which is the curvature. And thirdly it is, in fact, no longer true to say that all physically interesting fluids move at non-relativistic speeds. For example, near the horizon of a black hole, or near the surface of a neutron star, or near the centre of an atomic bomb, gases move under conditions so extreme that relativistic effects *do* become important.

As one might expect, there are many routes to the relevant equations. One is to *postulate* them, and then show that (i) they conform to the various conservation laws which by now we have come to expect and (ii) they reduce to the already well-tested laws of relativistic particle mechanics when the fluid becomes particles.

Another approach, which in fact we shall follow here, is to *construct* the equations heuristically first, and *then* postulate them.

We shall assume that a material fluid has at each event a *unique* 'rest frame' S_0, i.e. an inertial frame in which its momentum density at that event momentarily vanishes. This corresponds in particle mechanics to the fact that the three-momentum of a system vanishes in a unique frame S_{CM}. [For an electromagnetic field the assumption would be false: a null field has *no* rest frame, and a non-null field has many.] Now from our experience with the electromagnetic energy tensor it would seem reasonable to postulate that the energy tensor $T^{\mu\nu}$ of a material fluid will have the following components *in the rest frame*:

$$T^{\mu\nu} = \begin{pmatrix} c^2\rho_0 & 0 & 0 & 0 \\ 0 & t_0^{11} & t_0^{12} & t_0^{13} \\ 0 & t_0^{21} & t_0^{22} & t_0^{23} \\ 0 & t_0^{31} & t_0^{32} & t_0^{33} \end{pmatrix} \tag{44.1}$$

where $c^2\rho_0$ is the energy density and t_0^{ij} is the usual Newtonian stress tensor—for, after all, we also know that for *slow* motions Newtonian mechanics becomes relevant. Equation (44.1) then defines the components of $T^{\mu\nu}$ in all inertial frames. The only problem is to *recognize* them. For example, in a frame S in standard configuration with S_0, we find $T^{00} = \gamma^2(c^2\rho_0 + c^{-2}v^2t_0^{11})$. What is the physical meaning of this? Is it satisfactory simply to *assert* that this is the energy density in S without trying to understand why? It is for reasons such as these that we shall here follow the constructive rather than the axiomatic approach. We shall let ourselves be guided by the assumption that the conservation laws of energy, momentum, and angular momentum can be extended to arbitrary fluids.

45. Preliminaries. External and internal forces

We shall make our calculations in terms of the usual coordinates ct, x, y, z(or x^0, x^1, x^2, x^3) of some definite inertial frame S. We shall generally be interested in the behaviour of the fluid at some particular event, and we shall denote by **u** its velocity at that event (i.e. the velocity of its rest frame in which its momentum density vanishes). We can always re-orient the spatial axes in S, if necessary, so that **u** points along the positive x-axis. This system of coordinates we shall denote by $S_\parallel$. The instantaneous rest frame of the fluid at the event of interest, in standard configuration with $S_\parallel$, will be denoted by S_0. All quantities measured in S_0 will be distinguished by the subscript 0. Thus, for example, ρ will denote the mass density in S and ρ_0 the proper mass density, i.e. the density as measured in S_0.

Two kinds of time derivative must be distinguished: the *local* derivative, $\partial/\partial t$, giving the rate of change of a quantity at a point fixed in the frame, and the *comoving* or *Eulerian* derivative, d/dt, giving the rate of change at a point fixed in the fluid. Any quantity $Q(ct, x, y, z)$ associated with the fluid as a whole can be specialized to one specific point $\mathbf{r} = \mathbf{r}(t)$ fixed in the fluid; its comoving derivative is then given by

$$\frac{dQ}{dt} = \frac{\partial Q}{\partial x^\mu}\frac{dx^\mu}{dt} = \frac{\partial Q}{\partial t} + u^i\frac{\partial Q}{\partial x^i}. \tag{45.1}$$

We shall later need a formula for the (Eulerian) rate of change of the volume of a given element of fluid. Consider, therefore, a small rectangular element having volume dV and edges dx^1, dx^2, dx^3 parallel to the corresponding coordinate axes. In time dt one face perpendicular to the x^1-axis sweeps out a volume $u^1 dt dx^2 dx^3$, whereas the other sweeps out a volume

$$\left(u^1 + \frac{\partial u^1}{\partial x^1}dx^1\right)dt\, dx^2 dx^3.$$

Hence the change in volume due to this pair of faces is $(\partial u^1/\partial x^1)dV dt$, and similar expressions hold for the other two pairs. The total rate of change of dV is therefore given by

$$\frac{d}{dt}(dV) = \frac{\partial u^i}{\partial x^i}dV. \tag{45.2}$$

For future reference we also note that the choice of a unit of length is arbitrary; thus the 'unit areas' and 'unit volumes' which we shall sometimes employ to shorten the writing out of certain arguments are, in fact, arbitrarily small. The alternative would be to carry factors of type $dx dy$, $dx dy dz$, through the equations.

Now a fluid is generally subject to two types of forces: external and internal. The former are often called 'volume' forces, because their effect is locally proportional to the size of the (infinitesimal) volume on which they act. Examples are provided by electromagnetic forces acting on a continuous distribution of charge, or gravitational forces acting on a continuous distribution of mass. We shall write $\mathbf{k}$ and $\tilde{\mathbf{k}}$ for the external three-force per unit volume and per unit proper volume, respectively, and $\tilde{K}^\mu$ for the four-force corresponding to $\tilde{\mathbf{k}}$. Then if, as we shall assume, the external forces are 'pure', we have, as in (42.12),

$$\tilde{K}^\mu = \gamma(u)(\tilde{\mathbf{k}} \cdot \mathbf{u}/c, \tilde{\mathbf{k}}) = (\mathbf{k} \cdot \mathbf{u}/c, \mathbf{k}). \tag{45.3}$$

The other type of forces which we must take into account are the internal forces, i.e. the elastic stresses within the fluid itself. Consider a small 'test' area *moving with the fluid*. The matter on either side of this area will experience elastic forces, which: (i) will be equal and opposite, by Newton's third law (cf. Section 35); (ii) will not necessarily be normal to the test area, since a general (viscous) fluid can sustain both normal and tangential stresses; and (iii) will evidently be locally proportional to the size of the (infinitesimal) test area, whence these forces are often called 'area' forces. We denote by $\mathbf{t}^{(j)}$ the three-force (measured in S) exerted across a unit area fixed in S_0 perpendicularly to the x^j-axis, by the fluid on the negative side of this axis upon the fluid on the positive side. The i component of $\mathbf{t}^{(j)}$ we denote by t^{ij}. Then we find, by an argument similar to that for (42.16), that the total elastic force acting on a unit volume of fluid is given by

$$f^i = -t^{ij}{}_{,j}. \tag{45.4}$$

The components t^{ij} constitute, as we shall see presently, a three-tensor which is called the *elastic stress tensor*. But first we show that a knowledge of the t^{ij} at any given point enables us to write down the stress on an *arbitrarily* oriented test area moving with the fluid at that point. For consider a small fluid tetrahedron with three of its faces, of area dA_1, dA_2, dA_3, perpendicular to the coordinate directions $\mathbf{i}_1, \mathbf{i}_2, \mathbf{i}_3$, respectively (see Fig. 20). Let the fourth face, of area dA, have unit

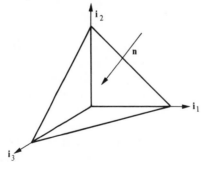

FIG.   20

inward normal $\mathbf{n} = (n_1, n_2, n_3)$, and experience an elastic force $\mathbf{t}$ $= (t^1, t^2, t^3)$ per unit area due to the external fluid. Since the areas $\mathrm{d}A_j$ are the projections of $\mathrm{d}A$ onto the respective coordinate planes, we have

$$\mathrm{d}A_j = -\mathrm{d}A\,\mathbf{n}\cdot\mathbf{i}_j = -\mathrm{d}A n_j. \tag{45.5}$$

Also the volume of the tetrahedron is given by

$$\mathrm{d}V = \tfrac{1}{3}\mathrm{d}A h, \tag{45.6}$$

where h is the altitude on the face $\mathrm{d}A$. Consider now the equation of motion (35.4) as applied to this portion of fluid,

$$\mathbf{t}^{(1)}\mathrm{d}A_1 + \mathbf{t}^{(2)}\mathrm{d}A_2 + \mathbf{t}^{(3)}\mathrm{d}A_3 + \mathbf{t}\mathrm{d}A + \mathbf{k}\mathrm{d}V = \frac{\mathrm{d}}{\mathrm{d}t}(\mathbf{g}\mathrm{d}V)$$
$$= \mathrm{d}V\left(\frac{\mathrm{d}\mathbf{g}}{\mathrm{d}t} + \mathbf{g}\frac{\partial u^i}{\partial x^i}\right),$$

where $\mathbf{g}$ is the momentum density of the fluid, and where we have used (45.2). If we divide by $\mathrm{d}A$ and let $h \to 0$ while keeping the tetrahedron similar to itself, we find, using (45.5) and (45.6), the following expression for $\mathbf{t}$:

$$\mathbf{t} = \mathbf{t}^{(j)}n_j, \text{ i.e. } t^i = t^{ij}n_j. \tag{45.7}$$

The tensor property of t^{ij} now follows from the quotient rule, since t^i and n_j evidently are three-vectors.

In classical theory the t^{ij} can be shown to be symmetric, $t^{ij} = t^{ji}$, in all inertial frames. (In fact, being force components, they are left unchanged by a Galilean transformation.) In relativity this is not generally true [cf.(45.9) below], but it *is* true in the instantaneous rest frame, i.e.

$$t_0^{ij} = t_0^{ji}. \tag{45.8}$$

This symmetry is of more than casual interest: without it, Einstein's field equations of general relativity in their present natural form would be impossible. Furthermore, it is connected with the equivalence of mass and energy (see penultimate paragraph of Section 42) and also with the conservation of angular momentum. The following standard proof, in fact, shows that last connection. Consider the motion of a small cube of fluid relative to the instantaneous rest frame S_0 of its centre P. Let the edges of the cube be of length $\mathrm{d}l$ and parallel to the coordinate axes. The velocities of all parts of the cube relative to

S_0 will be infinitesimal and therefore Newtonian arguments apply. We shall assume that the resultant elastic force on any face of the cube acts at the centre of that face. Now let us take moments about an axis through P which is parallel to the x^h-axis. If t_0^{ij} (h, i, j all different) refers to P, then we have, by applying the principle of angular momentum to the cube in S_0,

$$\pm\,[t_0^{ij} - t_0^{ji} + O(\mathrm{d}l)]\mathrm{d}l^3 = \text{(rate of change of angular momentum)} - \text{(moment of external forces)}.$$

In this equation $O(\mathrm{d}l)$ denotes a quantity of the order of magnitude of $\mathrm{d}l$ and appears because of the possible deviation of t_0^{ij} and t_0^{ji} at the faces from their values at P. Evidently the two terms on the right are $O(\mathrm{d}l^4)$; thus when we divide the equation by $\mathrm{d}l^3$ and let $\mathrm{d}l \to 0$ we obtain the identity (45.8).

As we shall presently need the relation between the stress components in S_0 and $S_\parallel$, let us get this next. Consider, for example, the x-component of the force on an area $\mathrm{d}A$ perpendicular to the y-axis. This force is $t^{12}\mathrm{d}A$ in $S_\parallel$ and $t_0^{12}\mathrm{d}A_0$ in S_0. But, by (35.6), these forces are equal, and, by length contraction, $\mathrm{d}A = \mathrm{d}A_0/\gamma(u)$. Consequently $t^{12} = t_0^{12}\gamma(u)$. In the same way we obtain the remaining equations in the following array:

$$
\begin{aligned}
&t^{11} = t_0^{11}, && t^{12} = t_0^{12}\gamma(u), && t^{13} = t_0^{13}\gamma(u) \\
&t^{21} = t_0^{21}/\gamma(u), && t^{22} = t_0^{22}, && t^{23} = t_0^{23}, && (45.9) \\
&t^{31} = t_0^{31}/\gamma(u), && t^{32} = t_0^{32}, && t^{33} = t_0^{33}.
\end{aligned}
$$

This in conjunction with (45.8) bears out our assertion that t^{ij} is not generally symmetric.

46. The augmented momentum and mass densities

We now turn our attention to two somewhat subtle and purely relativistic effects. The first concerns the fact that even a non-material energy flow, being equivalent to a flow of mass, has a momentum of its own which must be added to the material momentum in all relevant equations. For example, if I accelerate a massive object by pulling it with a rope towards me, the rope will have material momentum in a direction towards me. But also, since it is transmitting energy ($\equiv$ mass) from me to the object, there will be an immaterial momentum along the rope in the opposite direction.

Consider a unit area moving with the fluid and perpendicular to the x^i-axis. The matter on either side of this area experiences a force, which is $\mathbf{t}^{(i)}$ on one side and $-\mathbf{t}^{(i)}$ on the other. Since the area moves with the fluid, these forces do work at rates $\pm \mathbf{u} \cdot \mathbf{t}^{(i)}$ respectively. Thus the fluid on one side of the area gains energy while the fluid on the other side loses energy, and the rate at which (immaterial) energy crosses this unit area in the direction of the positive x^i-axis is $\mathbf{u} \cdot \mathbf{t}^{(i)}$ $= u_j t^{ji}$. As we have seen in (42.14), such an energy current is equivalent to a momentum density $u_j t^{ji}/c^2$. The momentum density due to the moving matter (including, of course, all energy stored within it elastically or otherwise) is ρu^i, whence the *total momentum density* g^i is given by

$$g^i = \rho u^i + u_j t^{ji}/c^2. \tag{46.1}$$

In the present analysis we assume that there is no conduction of heat in the fluid. If there were, a corresponding mass current would yet have to be added to the right-hand side of (46.1) [and to (44.1)].

It may be objected that, in the above argument, $u_j t^{ji}$ represents the energy current across an area which *moves* relative to the reference frame and that it should be corrected for that motion. But this cannot be done, since in general we cannot ascribe a unique velocity to an immaterial energy current [cf. Exercise VI(13)]. Nor *need* it be done. Consider, for example, a flat piston pushing with force f, (non-relativistic) velocity u, and acceleration a, a square block of mass m. The rate at which the block gains energy is $fu = mau$. Now suppose a plane fixed in the reference frame coincides momentarily with the plane interface between piston and block. Clearly the immaterial energy transfer across *either* plane, to first order, is $maudt$ in time dt.

The second relativistic effect to be discussed here concerns the mass density ρ. It would seem at first sight that ρ and ρ_0 should be related by the equation $\rho = \gamma^2(u)\rho_0$, where one γ is due to length contraction affecting what is a unit volume in the rest frame, and the other is due to mass increase according to formula (26.3). But that simple formula is valid only in certain special cases, e.g. for single particles and for systems of *free* particles [cf. (30.4)]. It is *not* generally valid for constrained systems. This can be seen in the following way. In S_0 consider a volume element of fluid which is a cube with its edges, of length dl, parallel to the coordinate axes. Suppose we could instantaneously remove all the fluid surrounding this cube. The total energy of the cube would not be altered by this, and hence its total mass, which is $\rho_0 dl^3$, would not change. But now look at this situation in $S_\parallel$. Here the fluid surrounding the cube has *not* been removed simultaneously. The 'left' face (open to the negative x-axis) has been exposed $\gamma(u)udl/c^2$ seconds before the opposite (right) face, as follows from the Δ version of (7.4) (i). During that length of time the material of the cube has done work on the fluid adjacent to the right face at the rate $ut^{11}dl^2$, without receiving a corresponding amount of energy from the left. The net work done by the cube on the fluid adjacent to the other faces depends on the *variation* of t^{ij} and so involves one more power of dl; in the limit as $dl \to 0$ it therefore becomes negligible. Thus altogether the cube has lost $\gamma(u)u^2t^{11}dl^3/c^2$ units of energy, which corresponds to $1/c^2$ times that much mass, in the process. When the surrounding fluid has been completely removed the cube can be regarded as a compound particle, and hence its total mass in $S_\parallel$ will be $\gamma(u)\rho_0 dl^3$. Thus before the removal it must have been $[\gamma(u)\rho_0 + \gamma(u)u^2t^{11}/c^4]dl^3$. If we now take account of the length contraction of the cube, and also use (45.9) (i), we finally obtain the mass density of the constrained fluid in $S_\parallel$ in terms of quantities measured in S_0:

$$\rho = \gamma^2(u)\left(\rho_0 + \frac{u^2}{c^4}t_0^{11}\right). \tag{46.2}$$

[Compare this with the expression for T^{00} at the end of Section 44.]

We can now also find expressions for the momentum density in $S_\parallel$ in terms of quantities measured in S_0, which will be needed presently. Substituting from (45.9) and (46.2) into (46.1), and setting $u^i = (u, 0, 0)$ for our case, we obtain, after a simple reduction,

$$\begin{aligned}
g^1 &= u\gamma^2(u)(\rho_0 + t_0^{11}/c^2), \\
g^2 &= u\gamma(u)t_0^{12}/c^2, \\
g^3 &= u\gamma(u)t_0^{13}/c^2.
\end{aligned} \tag{46.3}$$

47. The equations of continuity and of motion

We shall now apply the principles of energy and momentum to the fluid under consideration. Let us consider the energy principle first.

The total energy inside a closed boundary fixed in the frame of reference can increase in two ways: (i) energy (mass) can flow in through the boundary; and (ii) the external forces can do work on the fluid within. In connection with (i) we recall the relation (42.14) between energy current ε and momentum density $\mathbf{g}$. Thus the rate at which energy enters a *unit volume* through the boundary is given by. $-\operatorname{div} \varepsilon = -c^2 \operatorname{div} \mathbf{g}$. The rate at which the external forces do work on the fluid inside is $\mathbf{k} \cdot \mathbf{u}$. Together these quantities must equal $\partial(c^2 \rho)/\partial t$, the rate at which the energy inside increases. This yields

$$\frac{\partial \rho}{\partial t} = -\operatorname{div} \mathbf{g} + \frac{1}{c^2} \mathbf{k} \cdot \mathbf{u}, \tag{47.1}$$

the so-called *equation of continuity* of energy [cf. (42.13)].

Next let us apply the momentum principle (35.4) to a volume element $\mathrm{d}V$ of the fluid. By the definition of g^i and k^i, the momentum of this element and the external force on it are, respectively, $g^i \mathrm{d}V$ and $k^i \mathrm{d}V$, while the force exerted on it by the elastic stresses is, by (45.4), $-t^{ij}{}_{,j} \mathrm{d}V$. Consequently we have

$$\frac{\mathrm{d}}{\mathrm{d}t}(g^i \mathrm{d}V) = \left(k^i - \frac{\partial t^{ij}}{\partial x^j} \right) \mathrm{d}V. \tag{47.2}$$

Now, by use of (45.1) and (45.2),

$$\begin{aligned}
\frac{\mathrm{d}}{\mathrm{d}t}(g^i \mathrm{d}V) &= \frac{\mathrm{d}g^i}{\mathrm{d}t} \mathrm{d}V + g^i \frac{\mathrm{d}}{\mathrm{d}t}(\mathrm{d}V) \\
&= \left(\frac{\partial g^i}{\partial t} + \frac{\partial g^i}{\partial x^j} u^j \right) \mathrm{d}V + g^i \frac{\partial u^j}{\partial x^j} \mathrm{d}V,
\end{aligned}$$

whence (47.2) reduces to

$$\frac{\partial g^i}{\partial t} + \frac{\partial}{\partial x^j}(g^i u^j + t^{ij}) = k^i. \tag{47.3}$$

This is the *equation of motion*. Alternatively it can be interpreted as the *equation of continuity* of momentum [cf. (42.15)]. For $g^i u^j$ is the velocity-dependent momentum current, while t^{ij}, being a force, represents absorption of momentum, and thus an additional momentum current.

It is convenient at this stage to introduce new symbols p^{ij} (which will turn out to be symmetric) by the equations

$$p^{ij} = g^i u^j + t^{ij}, \tag{47.4}$$

in terms of which the equation of continuity (47.3) becomes

$$\frac{\partial g^i}{\partial t} + \frac{\partial p^{ij}}{\partial x^j} = k^i. \tag{47.5}$$

From their definition (47.4) in terms of known three-tensors it is evident that the p^{ij} also constitute a three-tensor. By the above discussion, this tensor is seen to represent the *total momentum current density*. Alternatively it is sometimes called the *total stress tensor*.

We shall now find expressions for p^{ij} in $S_\parallel$ in terms of quantities measured in S_0, which will be needed below. By substituting from (46.3) and (45.9) into (47.4), and setting $u^i = (u, 0, 0)$ as before, we find, after some simple reductions,

$$
\begin{aligned}
p^{11} &= \gamma^2(u)(t_0^{11} + \rho_0 u^2), & p^{12} &= \gamma(u)t_0^{12}, & p^{13} &= \gamma(u)t_0^{13}, \\
p^{21} &= \gamma(u)t_0^{21}, & p^{22} &= t_0^{22}, & p^{23} &= t_0^{23} \\
p^{31} &= \gamma(u)t_0^{31}, & p^{32} &= t_0^{32}, & p^{33} &= t_0^{33}.
\end{aligned}
\tag{47.6}
$$

Note, from (47.6) and (45.8), that p^{ij} is symmetric in $S_\parallel$; But since an arbitrary rotation of the spatial axes in $S_\parallel$ leaves the symmetry properties of a three-tensor unaltered, it follows that p^{ij} is symmetric in the general frame.

The quantities recognized as the 'stress' components of the electromagnetic field [cf. Section 42] are the analogues of the p^{ij} of this section. No analogues of the t^{ij} can, in general, be found for that field. For although the electromagnetic p^{ij} and g^i are determinate, the relevant u^i is *not* [cf. Exercise VI(13)], and therefore, by (47.4), the t^{ij} are also indeterminate.

48. The mechanical energy tensor

After the groundwork of the preceding sections we are now prepared to enunciate the axioms of relativistic fluid dynamics, and to find them reasonable. The energy tensor $T^{\mu\nu}$ of a fluid (called 'mechanical' to distinguish it from that of a field) is indeed defined by its 'Newtonian' components in the rest frame S_0 exactly as in our equation (44.1), and it is therefore certainly symmetric. When we now apply a standard (inverse) Lorentz transformation [cf. (22.6)–(22.8)]

$$T^{\mu\nu} = T^{\mu'\nu'}p_{\mu'}^\mu p_{\nu'}^\nu, \tag{48.1}$$

with $v = u$ to the energy tensor components $T^{\mu'\nu'}$ in S_0, we get the components $T^{\mu\nu}$ in $S_\parallel$. If we actually perform the simple computation indicated in (48.1) we find that

$$T^{\mu\nu} = \begin{pmatrix} c^2\rho & cg^1 & cg^2 & cg^3 \\ cg^1 & p^{11} & p^{12} & p^{13} \\ cg^2 & p^{21} & p^{22} & p^{23} \\ cg^3 & p^{31} & p^{32} & p^{33} \end{pmatrix}, \tag{48.2}$$

where ρ, g^i, p^{ij} are precisely the quantities appearing on the right-hand sides of our equations (46.2), (46.3), and (47.6). We are thus prepared to agree that, just as in the electromagnetic case, the mechanical energy tensor in *every* reference frame has for its time–time, time–space, and space–space parts the energy density, c times the momentum density, and the total stress tensor, respectively. For this is true in $S_\parallel$, and the general frame differs from $S_\parallel$ only by a possible rotation of the spatial axes, which preserves the various three-tensor parts of $T^{\mu\nu}$ [cf. Exercise IV(7)(ii)].

In the axiomatic approach, the above identification of the various parts of $T^{\mu\nu}$ is achieved *directly*, from the basic law postulated for fluid dynamics, which is

$$T^{\mu\nu}{}_{,\nu} = \tilde{K}^\mu, \tag{48.3}$$

where $\tilde{K}^\mu$ is the proper density of the external forces. With $\mu = 0$ this equation reads

$$\frac{\partial(c^2\rho)}{\partial t} + \operatorname{div}(c^2\mathbf{g}) = \mathbf{k}\cdot\mathbf{u}, \tag{48.4}$$

as is seen by reference to (48.2) and (45.3). We, of course, are prepared for this equation: it is our equation of continuity (47.1). But the axiomist too will regard (48.4) as the equation of continuity, $\mathbf{k}\cdot\mathbf{u}$ being the work done by the external forces. He is thus led to identify $c^2\rho$ as the energy density and $c^2\mathbf{g}$ as the energy current density, in all frames, just as we did in the corresponding electromagnetic case (42.13). He will, of course, also obtain the relations (46.2) and (46.3) by applying the Lorentz transformation (48.1), but he is then left to puzzle out their physical meaning.

The remaining information in (48.3) is obtained by putting $\mu = i$, whereupon it becomes

$$\frac{\partial g^i}{\partial t} + \frac{\partial p^{ij}}{\partial x^j} = k^i, \tag{48.5}$$

which is our equation of continuity (47.5). The axiomist, too, recognizes the momentum principle in (48.5), just as we did in the corresponding electromagnetic case (42.15). Thus equation (48.5) leads the axiomist to identify p^{ij} as the total momentum current. The only remaining problem to him will be to understand the physical significance of the relations (47.6).

The close analogy between the results of this section and those of Section 42 have been noted repeatedly. A remark from the penultimate paragraph there is also appropriate here: the symmetry of the energy tensor is intimately connected with Einstein's mass–energy equivalence.

When the fluid under consideration carries a continuous distribution of charge, and the only external volume forces acting on it are electromagnetic, then the $\tilde{K}^\mu$ of equation (48.3) is given by (42.1), which is equivalent to (42.4). Equation (48.3) can then be written in the form

$$(T^{\mu\nu} + M^{\mu\nu})_{,\nu} = 0. \tag{48.6}$$

If there are non-electromagnetic volume forces as well, only those will appear on the right-hand side of (48.6). This equation shows (for $\mu = 0$ and $\mu = i$, respectively) that for the combination fluid plus field both energy and momentum satisfy continuity equations in *every* region of the reference frame.

For 'ordinary' distributions of matter the energy density $c^2\rho$ *greatly* outweighs all the other components of $T^{\mu\nu}$, all of which have the dimensions of an energy density. Nevertheless $c^2\rho$ forms an inseparable whole with these other components, since every transformation from one inertial frame to another occasions a mingling of components.

Equation (46.2) makes it appear possible that for certain values of, say, t_0^{11}, ρ could become negative in some frames. This would occur when there are very strong *tensions* (negative t_0^{ii}). We recall that tensions, e.g. in the nucleus of an atom, represent 'binding' energy. In classical mechanics there are no limits to such negative binding energy (e.g. of two opposite point charges). Nevertheless $\rho < 0$ would lead to certain 'unphysical' situations such as the centre of mass of a fluid complex lying outside the complex. To avoid this, one usually assumes that $T^{\mu\nu}$ satisfies the so-called *weak energy condition* which requires

$$T^{\mu\nu} V_\mu V_\nu \geqslant 0$$

for all timelike vectors V_μ. By continuity this condition will then also hold for all null vectors. An observer having four-velocity U_μ measures an energy density $T^{\mu\nu} U_\mu U_\nu / c^2$ [since in his rest frame this invariant reduces to T^{00}]. Thus (48.7) is equivalent to the condition $\rho \geqslant 0$ in all inertial frames.

One logical step is still missing. We must show that when the fluid becomes particles, the laws of particle mechanics follow from those of fluid mechanics. For we cannot base *both* disciplines on arbitrary axioms when in fact they are clearly interrelated. But we shall postpone this demonstration until Section 50.

49. Perfect fluids and incoherent fluids

A fluid which cannot sustain tangential stress, i.e. which is completely non-viscous, is said to be *perfect*. A fluid which cannot sustain any stress whatever is said to be *incoherent*. As a sufficiently close approximation to an incoherent fluid one can picture a cloud of non-interacting dust particles. For perfect fluids and their subclass of incoherent fluids the energy tensor simplifies considerably, as we shall now show. These results are of importance, for example, in connection with general-relativistic cosmology.[1]

We shall first prove that in a perfect fluid the normal stress, or pressure, is the same in all directions. Consider equation (45.7), which in the present case $(t^{ij} = 0, i \neq j)$ can be written

$$p\mathbf{n} = t^{11}\mathbf{i}_1 n_1 + t^{22}\mathbf{i}_2 n_2 + t^{33}\mathbf{i}_3 n_3,$$

p being the pressure in the (arbitrary) direction $\mathbf{n}$. If we take the scalar product of both sides of this equation with $\mathbf{i}_1, \mathbf{i}_2, \mathbf{i}_3$ successively we find

$$p = t^{11} = t^{22} = t^{33}, \tag{49.1}$$

and this proves our assertion.

Another important fact is that p is *invariant* under transformation from one inertial frame to another. For consider any inertial frame S and any event $\mathscr{P}$ and the rest frame S_0 at $\mathscr{P}$. In S let dA be an area moving with the fluid at $\mathscr{P}$ and at right angles to the motion of the fluid. In S the elastic force on that area is pdA and in S_0 it is $p_0 dA_0$. But, by the remarks following (35.6), these forces are equal, and also $dA = dA_0$. It follows that

$$p = p_0, \tag{49.2}$$

bearing out our assertion. We could also have deduced this result from (45.9)(i) and (49.1).

Reference to (44.1) now shows that for a perfect fluid

$$T_0^{\mu\nu} = \text{diag}\,(c^2 \rho_0, p, p, p). \tag{49.3}$$

We assert that in the general frame $T^{\mu\nu}$ is then given by the tensor expression

$$T^{\mu\nu} = \left(\rho_0 + \frac{p}{c^2}\right) U^\mu U^\nu - g^{\mu\nu} p, \tag{49.4}$$

where $U^\mu = \gamma(u)\,(c, \mathbf{u})$ is the four-velocity of the fluid. This equation, by reference to (49.3), is seen to be satisfied in S_0, where $\mathbf{u} = 0$, and this establishes its general validity. In the case of an *incoherent* fluid ($p = 0$) it reduces to the very simple form

$$T^{\mu\nu} = \rho_0 U^\mu U^\nu. \tag{49.5}$$

We may note that a random distribution of radiation can also be regarded as a perfect fluid. This point of view can be justified by taking the distribution to be equivalent to a perfect gas of photons, but it also results from electromagnetic theory. For we have seen [in Exercise VI(17)] that the electromagnetic energy tensor for such a distribution in its 'rest frame' (the only frame in which it is random) is precisely of the form (49.3), with the extra condition

$$3p = \sigma_0 = c^2 \rho_0, \tag{49.6}$$

i.e. $T_\mu^\mu = 0$, which is *always* satisfied by an electromagnetic energy tensor [cf. Exercise VI(14)].

---

[1] There one studies the motion of the universe under its own gravity. One usually makes the assumption that this motion is not essentially altered when the actual universe is replaced by a uniform incoherent fluid having the same mean density ('dust universe'). Near the 'big bang', when the universe was much denser, one can allow for pressure by treating the universe as a perfect fluid.

50. Integral conservation laws

We have seen in Section 48 how the basic law (48.3) of fluid dynamics is equivalent to two equations of continuity, (48.4) and (48.5), one for energy and one for momentum. A continuity equation essentially asserts that whatever a small region *gains* (in energy, momentum, etc.)

the complementary outside region *loses*. It is therefore to be expected intuitively that such continuity equations imply global *conservation laws* of the relevant quantities. And this is indeed the first result we shall establish in this section.

As a mathematical tool we shall need the four-dimensional version of Gauss's familiar divergence theorem. We would expect this to have, and indeed it does have, the following form:

$$\int_R \xi^\mu{}_{,\mu} \, dR = \int_B \xi^\mu N_\mu \, dB, \tag{50.1}$$

where the integral on the left is actually a quadruple integral and that on the right is a triple integral; and where ξ^μ is any differentiable four-vector field defined over a region R of spacetime, dR is its four-dimensional volume element $\left| dx^0 \, dx^1 \, dx^2 \, dx^3 \right|$, B is the three-dimensional boundary of R, dB is *its* volume element, and N^μ is the unit outward normal to B. We shall not give a proof of this theorem here. But, in fact, the proof is quite analogous to the usual proof in three dimensions, except for some complications due to the indefiniteness of the spacetime metric.[1]

Now consider an isolated fluid complex K with energy tensor $T^{\mu\nu}$—such as a gas cloud or a drop of water floating in space—which is not subject to external forces. Relative to some specific inertial frame S, let us embed K at any instant $t = $ constant in a three-dimensional volume V whose boundary lies entirely in vacuum, and then define its instantaneous four-momentum P^μ by the integral

$$P^\mu = \frac{1}{c} \int_V T^{\mu 0} \, dV. \tag{50.2}$$

We shall presently prove not only that P^μ is constant in time, but also that, similarly defined in all other frames, it constitutes a four-vector! By substituting from (48.2) into (50.2) we see that this will mean that the total energy E and the total momentum $\mathbf{p}$ of K, defined by

$$E = \int_V c^2 \rho \, dV, \quad \mathbf{p} = \int_V \mathbf{g} \, dV, \tag{50.3}$$

are constant in time, and that they constitute a four-vector $\mathbf{P}$ thus:

$$\mathbf{P} = (E/c, \mathbf{p}). \tag{50.4}$$

To prove our first assertion, we shall apply Gauss's theorem (50.1) to the region R consisting of the 'world-tube' of the volume V between

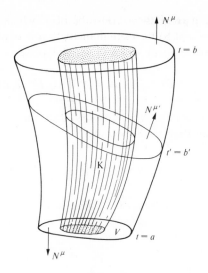

F IG.   21

two simultaneities $t = a$ and $t = b$ in S (see Fig. 21). On the mantle of this tube, which lies entirely in vacuum, we have $T^{\mu\nu} = 0$, while throughout the tube we have, by (48.3), $T^{\mu\nu}{}_{,\nu} = 0$. Let A_μ be an arbitrary *constant* vector field and take for ξ^μ in (50.1) the vector

$$\xi^\nu = A_\mu T^{\mu\nu}. \tag{50.5}$$

Then $\xi^\nu_{,\nu} = 0$ throughout R, whence the integral on the left of (50.1) vanishes. The integral on the right of (50.1) vanishes on the mantle, so the contributions to it from the end-faces must be equal and opposite:

$$A_\mu \int_{t=a} T^{\mu\nu} N_\nu \, dV = -A_\mu \int_{t=b} T^{\mu\nu} N_\nu \, dV. \tag{50.6}$$

But $N_\mu = \pm (1,0,0,0)$, respectively, on those faces, whence

$$A_\mu \int_{t=a} T^{\mu 0} \, dV = A_\mu \int_{t=b} T^{\mu 0} \, dV. \tag{50.7}$$

Finally we use the fact that A_μ is arbitrary: it implies that the two integrals in (50.7) must be equal. This establishes that P^μ is constant in time.

To prove that P^μ is a four-vector, we apply Gauss's theorem to the same vector ξ^μ, but over a somewhat different region R: this time let R

consist of that part of the world-tube of V which lies between a
simultaneity $t = a$ of S and a simultaneity $t' = b'$ of *another* frame S'
(see Fig. 21). Analogously to (50.6) we now obtain

$$A_\mu \int_{t=a} T^{\mu\nu} N_\nu \, dV = - A_\mu \int_{t'=b'} T^{\mu\nu} N_\nu \, dV. \qquad (50.8)$$

But $A_\mu T^{\mu\nu} N_\nu = A_{\mu'} T^{\mu'\nu'} N_{\nu'}$; $N_{\nu'} = (1,0,0,0)$ on $t' = b'$; and $A_{\mu'}$
$= A_\mu p^\mu_{\mu'}$; with all that, (50.8) becomes

$$A_\mu \int_{t=a} T^{\mu 0} \, dV = A_\mu p^\mu_{\mu'} \int_{t'=b'} T^{\mu'0'} \, dV.$$

Because of the arbitrariness of A_μ, this allows us to conclude that
$P^\mu = p^\mu_{\mu'} P^{\mu'}$, i.e. that P^μ is indeed a four-vector, and so both our
assertions are established.

The basic law (48.3) has implications also for the angular
momentum. Corresponding to the angular momentum four-tensor
$\bar{L}^{\mu\nu}$ of a free system of particles, we can define an angular momentum
four-tensor of an isolated fluid complex K relative to some origin-
event $\mathcal{O}$. We do this by the equation

$$L^{\mu\nu} = \frac{1}{c} \int_V (x^\mu T^{\nu 0} - x^\nu T^{\mu 0}) \, dV, \qquad (50.9)$$

where V and dV have the same significance as above, and where x^μ is
the position vector of a fluid element dV relative to $\mathcal{O}$. To prove that it
is constant in time we apply Gauss's theorem over the same region as
before, but this time to the vector

$$\xi^\sigma = A_{\mu\nu}(x^\mu T^{\nu\sigma} - x^\nu T^{\mu\sigma}), \qquad (50.10)$$

where $A_{\mu\nu}$ is an arbitrary constant tensor field. Then $\xi^\sigma = 0$ on the
mantle, and $\xi^\sigma{}_{,\sigma} = 0$ everywhere. [To see the latter, note that $x^\mu{}_{,\sigma}$
$= \delta^\mu_\sigma$.] Analogously to (50.7) we now obtain

$$A_{\mu\nu} \int_{t=a} (x^\mu T^{\nu 0} - x^\nu T^{\mu 0}) \, dV = A_{\mu\nu} \int_{t=b} (x^\mu T^{\nu 0} - x^\nu T^{\mu 0}) \, dV,$$

which again, because of the arbitrariness of $A_{\mu\nu}$, implies the equality
of the two integrals, and so the constancy of $L^{\mu\nu}$. Its tensor character
can be established in complete analogy to that of P^μ above.

As in Section 34, we can define the angular momentum three-vector

h of K by the equation [cf. end of Exercise IV(7)]

$$\mathbf{h} = (L^{23}, L^{31}, L^{12}),\tag{50.11}$$

for which we then find, from (50.9), (48.2), and the constancy of L^{ij},

$$\mathbf{h} = \int_V (\mathbf{r} \times \mathbf{g})\mathrm{d}V = \text{constant},\tag{50.12}$$

just as in Newtonian mechanics.

From the time constancy of L^{0i}, on the other hand, we can deduce—as in the case of a free system of particles—that the centroid of a free fluid complex moves with constant velocity. By reference to (50.9) and (50.3), we have

$$L^{0i} = \frac{1}{c}\int (ctT^{i0} - x^iT^{00})\mathrm{d}V$$

$$= ct\mathbf{p} - c\int \mathbf{r}\rho\mathrm{d}V = ct\mathbf{p} - c\mathbf{r}_C m,\tag{50.13}$$

where we have written m for the total mass $\int\rho\mathrm{d}V$ of K, and

$$\mathbf{r}_C = \int\mathbf{r}\rho\mathrm{d}V/\int\rho\mathrm{d}V\tag{50.14}$$

for the position vector of its centroid. Since L^{0i} is constant, differentiation of (50.13) yields the required result

$$\frac{\mathrm{d}\mathbf{r}_C}{\mathrm{d}t} = \frac{\mathbf{p}}{m},\tag{50.15}$$

in complete analogy to (34.8).

We are now ready for our final logical task [cf. end of Section 48], namely to show how the laws of particle mechanics follow from those of fluid mechanics. We shall need the following lemma. The centroid of a convex fluid complex K lies inside it if $\rho > 0$. For suppose this is not so. Then we can choose coordinates such that $\mathbf{r}_C = 0$ and K, being convex, lies entirely above the plane $z = 0$. But then

$$mz_C = 0 = \int z\rho\mathrm{d}V,$$

which is absurd, since the integrand is everywhere positive. This proves the lemma.

Now consider a single free particle as the limit of a very small convex fluid complex K. By our result (50.15), its centroid will move uniformly, and since the centroid is inside K, K itself will move uniformly. Moreover, (50.15) shows that the momentum **p** of the particle will be given by $m\mathbf{u}$, where m is its total mass and **u** is its velocity $d\mathbf{r}_C/dt$. Because $\mathbf{P} = (mc, m\mathbf{u})$ is then a four-vector [cf. (50.4)], we have, from equating its square in the general frame with that in the rest frame, our old relation $m = \gamma(u)m_0$ between the mass and the rest mass of a particle. And then we also have $\mathbf{P} = m_0\,\mathbf{U}$, **U** being the particle's four-velocity.

What if an external force F^μ acts on the particle? Assume that this force is approximately constant throughout the volume of the particle, and also throughout a small proper time interval $d\tau$ at the particle. Assume, too, that the particle has 'unit' proper volume [cf. after (45.2)], so that $F^\mu = \tilde{K}^\mu$. Now let us adapt the argument beginning with (50.5). From (48.3) we see that in the present case $\xi^\nu_{\ ,\nu} = A_\mu F^\mu$ inside the particle's world-tube. So when we apply Gauss's theorem (50.1) to a small portion of proper length $d\tau$ of that tube, the left-hand side becomes $A_\mu F^\mu d\tau$ (as we see by evaluating the invariant four-volume in the rest frame of the particle). The two sides of equation (50.7) will no longer be equal, but their *difference*, $A_\mu dP^\mu$, still equals the right-hand side of (50.1). Thus, 'cancelling' A_μ, we have derived our original equation of motion for particles, $F^\mu = dP^\mu/d\tau$, from the law of fluid mechanics.

The final law, that of the conservation of four-momentum in all particle collisions, is now almost self-evident. A particle collision, say one in which two particles go in and three come out, corresponds to a particular 'flow' pattern of fluid, in which two straight world-tubes coalesce and three straight world-tubes emerge. The time constancy of the total four-momentum P^μ of the fluid then immediately implies the same for the particles.

We may note in conclusion that our restriction of the discussion to 'pure' (i.e. rest mass preserving) external forces is quite inessential. One would clearly expect the basic axiom (48.3) to apply to all types of forces, and indeed it does. In the general case equation (48.4) would have $c\tilde{K}^0/\gamma$ in place of $\mathbf{k}\cdot\mathbf{u}$, which would independently lead to the identification of *that* term with the work done by the external forces in a unit volume of the reference frame.

---

[1] A detailed proof can be found in Synge (see note [2] at the end of Section 38, p. 120), Chapter VIII, Section 7. For historical reasons Synge calls this theorem 'Green's theorem'.

Exercises VII

1. A steel cable of proper density ρ_0 is subjected to a tension t per unit cross-sectional area. Find the maximum value of t for which $\rho > 0$ in all frames. (Actual steel cables break under a tension of about $200 \, \text{kg-wt/mm}^2$.)

2. From the form (49.5) of its energy tensor and the equation $T^{\mu\nu}{}_{,\nu} = 0$ prove directly that every portion of an incoherent fluid, subject to no external forces, moves with uniform velocity, i.e. that $A^{\mu} = dU^{\mu}/d\tau = 0$. [Hint: expand $\{(\rho_0 U^{\mu})U^{\mu}\}_{,\nu} = 0$ and use $U^{\mu}{}_{,\nu} U^{\nu} = A^{\mu}, U_{\mu} A^{\mu} = 0.$]

3. For much of relativistic cosmology, all one needs of relativistic fluid dynamics is the equation $(\rho_0 U^{\mu} U^{\nu})_{,\nu} = 0$ for a 'fluid' (representing an idealized universe) that consists of freely moving non-interacting particles, all having the same constant four-velocity U^{μ} at a given event. Prove this equation *ab initio* using only (45.2) and the conservation of mass.

4. The analogue for fluids of the 'rigid motion' of solids is 'incompressible motion': this is defined to be such that the proper volume of each fluid element is conserved. Prove that the necessary and sufficient condition for such motion is $U^{\mu}{}_{,\mu} = 0$. Prove also that, in the case of perfect fluids subject to pure external forces only, this condition is equivalent to $d\rho_0/dt = 0$, i.e. the constancy of ρ_0 for each fluid element.

5. For a slowly moving fluid (u^2/c^2 negligible) prove, by reference to (46.3) that

$$g^i = (\rho_0 \delta^{ij} + t_0^{ij}/c^2)u_j,$$

where δ^{ij} is the three-dimensional metric tensor.

6. Assume that, at some event, an energy tensor $T^{\mu\nu}$ satisfies $T^{\mu\nu} V_{\mu} V_{\nu} > 0$ for all timelike *and* null vectors V^{μ}. By reference to (43.13) show that for an electromagnetic energy tensor this inequality does not necessarily hold. But if it holds, prove that there is *at most* one 'rest frame' for $T^{\mu\nu}$, i.e. a frame in which $T^{0i} = 0$. [Hint: the four-velocity of the required frame satisfies $T^{\mu\nu} U_{\nu} = kU^{\mu}$ for some non-zero k (why?); if there were two such timelike 'eigenvectors' of $T^{\mu\nu}$ then there would also be a null eigenvector.] *Note*: one can also show that the inequality implies the existence of *at least* one rest frame, and, conversely, that a unique rest frame implies the inequality.

7. A rod of proper density ρ_0 and cross-sectional area A is at rest in a frame S, making an angle α with the x-axis, and subject to equal

and opposite external forces $\pm\mathbf{f}$ applied to its ends, in the direction of the rod. Find all components of the energy tensor in the rod.

8. Consider a stressed body which is stationary in some inertial frame. Then, when we differentiate (50.12) in the general frame we can take $\mathbf{g}\,\mathrm{d}V$ as constant (i.e. we attach the integration grid to the body) and so obtain $\mathrm{d}\mathbf{h}/\mathrm{d}t = \int(\mathbf{u}\times\mathbf{g})\mathrm{d}V$. In Newtonian mechanics this integral vanishes since $\mathbf{g}$ is parallel to $\mathbf{u}$, but in relativity, by (47.4), $u_1 g_2 - u_2 g_1 = t_{12} - t_{21}$, etc. and so the integral does *not* generally vanish. Now consider the rod of the preceding exercise, but view it from the usual second frame S'. Using (35.6), prove that here the rod and the applied forces are *not* parallel, and that consequently there is a couple $\mathbf{m}$ tending to turn the rod. Of course, the rod does not turn. That is because *without* its turning the equation $\mathbf{m} = \mathrm{d}\mathbf{h}/\mathrm{d}t$ is satisfied; prove this. [*Note*: an outline of a proof for the validity of the formula $\mathbf{m} = \mathrm{d}\mathbf{h}/\mathrm{d}t$ in relativity will be found in Exercise 10 below.]

9. Consider Lewis and Tolman's lever paradox. The pivot B of a right-angled lever ABC is fixed at the origin of S' while A and C lie on the positive x and y-axes, respectively, and $\mathrm{AB} = \mathrm{BC} = a$. Two numerically equal forces of magnitude f act at A and C, in directions BC and BA, respectively, so that equilibrium obtains. Show that in the usual second frame S there is a clockwise couple fav^2/c^2 due to these forces. Why does the lever not turn? Resolve the paradox along the lines of the preceding exercise. [*Note*: *Intuitively* it is not difficult to see how the angular momentum $\mathbf{h}$ of the lever continually increases in S in spite of its non-rotation. The force at C continually does positive work on the lever, while the reaction at B does equal negative work. Thus energy flows in at C and out at B. But an energy current corresponds to a momentum density. So there is a constant non-material momentum density along the limb BC towards B. It is the continual increase of *its* moment about the origin of S that causes $\mathrm{d}\mathbf{h}/\mathrm{d}t$ to be non-zero.]

10. A finite stressed body with angular momentum $\mathbf{h}$ is subject to external surface forces which constitute a couple $\mathbf{m}$. By use of the three-dimensional Gauss divergence theorem, convert the volume integral for $\mathrm{d}\mathbf{h}/\mathrm{d}t$ into an integral of the stresses t^{ij} over the surface of the body, and recognize this as $\mathbf{m}$. [*Hint*: fill in the details of the following proof, where equation numbers indicate results used.

$$\mathrm{d}h_1/\mathrm{d}t = \int [x_2 t^{3j},_j - x_3 t^{2j},_j + u_2 g_3 - u_3 g_2]\mathrm{d}V$$

$$(50.12),\ (47.2)$$

$$= \int [\ldots + t^{32} - t^{23}] \mathrm{d}V \qquad (47.4)$$

$$= \int \left[(x_2 t^{3j})_{,j} - (x_3 t^{2j})_{,j} \right] \mathrm{d}V$$

$$= \oint (x_2 t^{3j} n_j - x_3 t^{2j} n_j) \mathrm{d}S = m_1, \qquad (45.7)$$

and similarly for m_2 and m_3.]

APPENDIX
TENSORS FOR SPECIAL RELATIVITY

A1. Introduction

In this appendix we develop the elements of pure tensor theory to the extent required by special relativity. No use is made of tensors, however, before Chapter IV in this book.

Tensors are of great importance in connection with coordinate transformations. They serve to isolate intrinsic geometric and physical properties from those that merely depend on the coordinates. And since relativity is much concerned with coordinate transformations it is not surprising that tensors have been found to be the ideal mathematical tool for its study. *Special* relativity, in particular, is essentially concerned only with *linear* coordinate transformations. Consequently in what follows we shall stress certain results that apply particularly to the linear case.

A2. Preliminary description of tensors

Consider a set of N real variables $\{x^1, x^2, \ldots, x^N\}$, which we may write as $\{x^i\}$ ($i = 1, 2, \ldots, N$). It is useful, though not necessary, to regard these variables as coordinates in some N-dimensional space V_N, and to adopt a geometric language. Any non-singular transformation of the xs to a new set of variables $\{y^i\}$ can then be regarded as a re-coordinatization of V_N. We may associate with any such space V_N a system of tensors. Tensors are objects in V_N that can be *described* by their components, which form an ordered set of, say M, real numbers $A_1, \ldots, A_M$. (Think of the familiar 'three-vectors' of classical physics associated with Euclidean three-space E_3.) These components generally change when the coordinates of V_N are changed. In fact, we may define a tensor $\mathscr{A}$ as a *map* $\mathscr{A}: \Sigma \to \mathrm{R}_M$ from the set Σ of permissible coordinate systems $\{x^i\}$ to the space R_M of real-number M-tuplets $(A_1, \ldots, A_M)$. But not every such map is a tensor: to be one, it must satisfy certain conditions which will be elaborated below (see Section A5 and the end of Section A7). Like a three-vector, a tensor can be defined at *one* point of V_N only ('point-tensor'), in which case its components are just numbers, or it can be defined on some larger subspace of V_N ('field-tensor'), in which case its components are functions of position in V_N.

In N-dimensional space a tensor of *rank r* has $M = N^r$ components. In particular, a tensor of rank zero has one component A and is called a *scalar*. A tensor of rank one has N components $(A_1, \ldots, A_N)$ and is called a *vector*. A tensor of rank two has N^2 components, which can be exhibited in matrix form thus:

$$\begin{pmatrix} A_{11} & A_{12} & \ldots & A_{1N} \\ A_{21} & A_{22} & \ldots & A_{2N} \\ \ldots & \ldots & \ldots & \ldots \\ A_{N1} & A_{N2} & \ldots & A_{NN} \end{pmatrix}.$$

Such a tensor is said to be *singular* if the determinant of this matrix vanishes. It is said to be *symmetric* if $A_{ij} = A_{ji}$, or *anti-symmetric* if $A_{ij} = -A_{ji}$, for all relevant i and j. Tensors of higher rank cannot be exhibited in such convenient forms, but tensors of all ranks are usually represented by a typical component, e.g. we may loosely speak of 'the tensor A_{ijk}' (rank 3), 'the tensor A_{ijkl}' (rank 4), etc.[1] In general, the order of the indices is significant, e.g. $A_{132} \neq A_{321}$, though specific tensors of all ranks may possess various symmetry properties such as $A_{ijk} = A_{jki}$. The indices i, j, k, $\ldots$ will always be understood to range from 1 to N.

For reasons that will become apparent we shall use both super-scripts and subscripts in what follows, and the level of an index will be significant. Typical tensor components may look thus: A^{ij} (rank 2), B^i_j (rank 2), C^i_{jkl} (rank 4), etc. When we exhibit the components of a second-rank tensor $(A_{ij}, A^{ij}, \text{or } A^i_j)$ as a matrix, the first or upper index will always refer to the row and the other to the column. When the off-diagonal components vanish, we may write

$$A_{ij} = \text{diag}(A_{00}, A_{11}, \ldots, A_{NN}).$$

[1] A more logical notation would be (A_{ijk}) for the entire tensor and A_{ijk} for a single component; but this can become tedious and we shall not use it.

A3. The summation convention

We shall find it convenient to use *Einstein's summation convention*, namely: if any index appears twice in a given term, once as a subscript and once as a superscript, a summation over the range of that index is implied. Thus, for example,

$$A_i A^i = \sum_{i=1}^{N} A_i A^i = A_1 A^1 + A_2 A^2 + \ldots + A_N A^N;$$

$$\begin{aligned}
A_{ijk} B^{ij} = \sum_{i=1}^{N} \sum_{j=1}^{N} A_{ijk} B^{ij} = {} & A_{11k} B^{11} + A_{12k} B^{12} + \ldots + A_{1Nk} B^{1N} \\
& + A_{21k} B^{21} + \ldots \qquad\qquad + A_{2Nk} B^{2N} \\
& \qquad\qquad \ldots \\
& + A_{N1k} B^{N1} + \ldots \qquad\qquad + A_{NNk} B^{NN}.
\end{aligned}$$

By a slight extension of the rule we shall also understand summation in such expressions as

$$\frac{\partial u^i}{\partial x^i}, \quad \frac{\partial q}{\partial x^i} \frac{\mathrm{d} x^i}{\mathrm{d} t}, \text{ etc.}$$

In certain manipulations the reader will at first find it helpful to imagine the summation signs in front of the relevant terms: since the summations are all finite, all elementary rules, such as interchanging the order of summation, differentiating under the summation sign, etc. apply. The repeated indices signalling summation are called *dummy indices* while a non-repeated index is called a *free index*. An obvious but important principle is that a dummy index pair can be replaced by any other: e.g. $A_i A^i = A_j A^j$. Such a replacement is often necessary to avoid the triple occurrence of an index which might lead to ambiguities. On the rare occasions when we wish to suspend the summation convention we can use the symbol NS ('no summation'). Thus $A_i A^i(\text{NS})$ denotes a single term.

A4. Coordinate transformations

To distinguish between various coordinate systems for our space V_N we shall use primed and multiply primed indices. All range from 1 to N. Thus:

$$i, j, k, \ldots; i', j', k', \ldots; i'', j'', k'', \ldots; \ldots = 1, 2, \ldots, N.$$

No special relation is implied between, say, i and i': they are as independent as i and j'. A first system of coordinates can then be denoted by $\{x^i\} = \{x^1, x^2, \ldots, x^N\}$, a second by $\{x^{i'}\} = \{x^{1'}, x^{2'}, \ldots, x^{N'}\}$, etc. Similarly the components of a given tensor in different coordinate systems are distinguished by the primes on their indices. Thus, for example, the components of some third-rank tensor may be denoted by A_{ijk} in the $\{x^i\}$ system, by $A_{i'j'k'}$ in the $\{x^{i'}\}$ system, etc. When primed indices take particular numerical values, we can prime *these*, so as not to lose sight of the relevant coordinate system. Thus, for example, when $i' = 2, j' = 3, k' = 5$, $A_{i'j'k'}$ becomes $A_{2'3'5'}$. This will already have been noted for the case of the coordinates above. (However, sometimes we adopt the simpler device of priming the kernel: A'_{235}.)

When we make a coordinate transformation from one set of coordinates x^i to another $x^{i'}$, it will be assumed that the transformation is non-singular, i.e. that the equations which express the $x^{i'}$ in terms of the x^i can be solved uniquely for the x^i in terms of the $x^{i'}$. We also assume that the functions specifying a transformation are differentiable as often as may be required. For convenience we write

$$\frac{\partial x^{i'}}{\partial x^i} = p_i^{i'}, \quad \frac{\partial x^i}{\partial x^{i'}} = p_{i'}^i, \quad \frac{\partial^2 x^{i'}}{\partial x^i \partial x^j} = p_{ij}^{i'} \qquad (A.1)$$

(p for 'partial derivative'), and use a similar notation for other such derivatives.

We observe that, by the chain rule of differentiation,

$$p_{i'}^i p_{i''}^{i'} = p_{i''}^i, \quad p_{i'}^i p_j^{i'} = \delta_j^i, \qquad (A.2)$$

where δ_j^i (the *Kronecker delta*) equals 1 or 0 according as $i = j$ or $i \neq j$. It is important to note the 'index-substitution' action of δ_j^i exemplified by $A_{ikl} \delta_j^i = A_{jkl}$.

A5. Informal definition of tensors

We are now ready to give an informal definition of tensors, which allows us to recognize a tensor when we see one. A mathematically formal definition of tensors is given at the end of Section A7.

(i) An object having components $A^{ij \cdots n}$ in the x^i system of coordinates and $A^{i'j' \cdots n'}$ in the $x^{i'}$ system is said to behave as a *contravariant* tensor under the transformation $\{x^i\} \to \{x^{i'}\}$ if

$$A^{i'j' \cdots n'} = A^{ij \cdots n} p_i^{i'} p_j^{j'} \cdots p_n^{n'} . \qquad (A.3)$$

(ii) Similarly, $A_{ij \ldots n}$ is said to behave as a *covariant* tensor under $\{x^i\} \to \{x^{i'}\}$ if

$$A_{i'j' \ldots n'} = A_{ij \ldots n} p_{i'}^i p_{j'}^j \cdots p_{n'}^n . \qquad (A.4)$$

(iii) Lastly, $A^{i \ldots k}_{l \ldots n}$ is said to behave as a *mixed* tensor (contravariant in $i \ldots k$ and covariant in $l \ldots n$) under $\{x^i\} \to \{x^{i'}\}$ if

$$A^{i' \ldots k'}_{l' \ldots n'} = A^{i \ldots k}_{l \ldots n} \, p^{i'}_i \ldots p^{k'}_k \, p^l_{l'} \ldots p^n_{n'}. \qquad (A.5)$$

Note that (A.5) evidently subsumes both (A.3) and (A.4) as special cases. The reader should perhaps be reminded that there are r separate summations going on in the above formulae if the tensors are of rank r.

At a given point in V_N the ps are pure numbers. Thus the tensor transformations (A.3)–(A.5) are linear: the components in the new coordinate system are linear functions of the components in the old system, the coefficients being products of the ps. Contravariant tensors involve derivatives of the new coordinates $x^{i'}$ with respect to the old, x^i, covariant tensors involve the derivatives of the old coordinates with respect to the new, and mixed tensors involve both types of derivatives. The convention of using subscripts for covariance and superscripts for contravariance, together with the requirement that the free indices on both sides of the equations must balance, serve as a perfect mnemonic for reproducing equations (A.3)–(A.5).

If we simply say an object *is* a tensor it is understood that the object *behaves* as a tensor under *all* non-singular differentiable transformations of the coordinates of V_N. An object which behaves as a tensor only under a certain subgroup of non-singular differentiable coordinate transformations may be called a 'qualified tensor', and its name should be qualified by an adjective recalling the subgroup in question, as in 'Lorentz tensor', more commonly called 'four-tensor'. These tensors are, as a matter of fact, the (qualified) tensors used in special relativity. [Exercise A(2) illustrates the concept of a qualified tensor under a different transformation group, the orthogonal group.] But we shall occasionally lapse from this strict terminology by omitting the adjective 'qualified' when no confusion seems likely.

The above definitions, when applied to a tensor of rank zero (a scalar) imply $A' = A$ (no ps!), whence a scalar is a function of position in V_N only, i.e. it is independent of the coordinate system. A scalar is therefore often called an *invariant*.

The *zero tensor* of any type $A^{i \ldots k}_{j \ldots n}$ is defined as having all its components zero in all coordinate systems. It is clear from (A.5) that it *is* a tensor. For brevity it is usually written as 0, with the indices omitted.

Evidently we must call two tensors *equal* if they constitute the same map $\Sigma \to R_M$(cf. Section A2), in other words if they have the same components in all relevant coordinate systems. Now the *main theorem* of the tensor calculus—trivial in its proof, profound in its implications—is this: if two tensors of the same type have equal components in *any one* coordinate system then they are equal. This is an immediate consequence of the definition (A.5). It implies that tensor-component equations always express physical or geometric facts, i.e. facts transcending the coordinate system used to describe them.

A6.   Examples of tensors

The simplest example of a *contravariant* vector is provided by the differentials of the coordinates, dx^i. For,

$$dx^{i'} = \frac{\partial x^{i'}}{\partial x^i}\, dx^i = dx^i\, p_i^{i'} .$$

(It is for this reason that we conventionally write coordinates with superscripts.) Under a *linear* transformation group the coordinate differences Δx^i transform like the differentials dx^i [cf. above (6.9)] and thus constitute a 'qualified' vector—usually called the *displacement vector*. Because of this the displacement vector can then serve to represent any contravariant vector. (Recall the 'directed line segments' of elementary vector analysis!) The coordinates x^i themselves behave as vectors only under linear homogeneous transformations [cf. above (6.9)]. A case in point is the homogeneous Lorentz transformation group.

The simplest example of a *covariant* vector is provided by the gradient of a function of position $\phi = \phi(x^1, \ldots, x^N)$. For, if we write

$$\phi_{,i} = \frac{\partial \phi}{\partial x^i} ,$$

we have

$$\phi_{,i'} = \frac{\partial \phi}{\partial x^{i'}} = \frac{\partial \phi}{\partial x^i}\frac{\partial x^i}{\partial x^{i'}} = \phi_{,i}\, p_{i'}^{i} .$$

An important example of a mixed second-rank tensor is provided by the Kronecker delta introduced after (A.2) above. For, by use of an analogue of (A.2), we have

$$\delta_j^i\, p_i^{i'}\, p_{j'}^{j} = p_j^{i'}\, p_{j'}^{j} = \delta_{j'}^{i'} .$$

A7. The group properties. Formal definition of tensors

It follows readily from their definitions that the tensor component transformations (A.5) satisfy the two *group properties* of *symmetry* and *transitivity* [cf. Section 7(vii)]. In other words, if an object behaves as a tensor under $\{x^i\} \rightarrow \{x^{i'}\}$ then it also behaves so under $\{x^{i'}\} \rightarrow \{x^i\}$ (symmetry); and if an object behaves as a tensor under $\{x^i\} \rightarrow \{x^{i'}\}$ and under $\{x^{i'}\} \rightarrow \{x^{i''}\}$ then it also behaves so under $\{x^i\} \rightarrow \{x^{i''}\}$ (transitivity). The general method of proof will be sufficiently indicated by dealing with an object of type A^i_j. Note the use of substitute dummies h and k in the first of the following equations to avoid the triple occurrence of i and j, and the use of (A.2) (or analogues of it):

$$A^i_{j'} p^i_{i'} p^{j'}_j = (A^h_k p^{i'}_h p^k_{j'}) p^i_{i'} p^{j'}_j = A^h_k \delta^i_h \delta^k_j = A^i_j ;$$

$$A^{i'}_{j''} = A^{i'}_{j'} p^{i''}_{i'} p^{j'}_{j''} = (A^i_j p^{i'}_i p^j_{j'}) p^{i''}_{i'} p^{j'}_{j''} = A^i_j p^{i''}_i p^j_{j''} .$$

As a result of the group properties, we can construct a tensor by specifying its components arbitrarily in *one* coordinate system, say $\{x^i\}$, and then using the transformation law (A.5) to define its components in all other systems. The group properties ensure that any two sets of components will then be related tensorially. For if A_i, say, is related tensorially to $A_{i'}$ and to $A_{i''}$, then $A_{i'}$ is so related to A_i (by symmetry) and consequently to $A_{i''}$ (by transitivity).

Even more importantly, the group properties allow us to make precise and formal our preliminary definitions of tensors given in Sections A2 and A5. Note that the permissible coordinate systems $\{x^i\}$ on V_N form an equivalence class, two such systems being equivalent if the transformation from the one to the other is non-singular, i.e. invertible. (In the case of a qualified tensor under a specific transformation group, any two coordinate systems related by a member of that group are to be regarded as equivalent.) Now, because of the group properties, if a tensor $\mathscr{A}$ associates the set of components $A^i_{k} \cdots$ with the system $\{x^i\}$, the 'pairs' $[\{x^i\}, A^i_{k} \cdots]$ also form an equivalence class, two such pairs being equivalent if the coordinate systems are equivalent, *and* if the corresponding As are tensorially related according to (A.5). *The tensor $\mathscr{A}$ can then be defined as the equivalence class of all these pairs.*

A8. Tensor algebra

The algebra of tensors consists of four basic operations—*sum, outer product, contraction,* and *index permutation*—which all have the property of producing tensors from tensors. All can be defined by the relevant operations on the tensor components, but must then be checked for tensor character.

The *sum* $C_k^i \cdots$ of two tensors $A_k^i \cdots$ and $B_k^i \cdots$ of the same *valence* [we shall say a tensor has valence (s, t) if its components have s contravariant and t covariant indices] is defined thus:

$$C_k^i \cdots = A_k^i \cdots + B_k^i \cdots .$$

Trivially it is a tensor (we exhibit the proof for a particular case):

$$C_{k'}^{i'} = A_{k'}^{i'} + B_{k'}^{i'} = A_k^i p_i^{i'} p_{k'}^k + B_k^i p_i^{i'} p_{k'}^k = (A_k^i + B_k^i) p_i^{i'} p_{k'}^k .$$
$$= C_k^i p_i^{i'} p_{k'}^k .$$

Note, however, that the sum of tensors at *different* points of V_N is not generally a tensor since in the third step above we could not generally pull out the ps. (It is this which complicates the concepts of derivative and integral in tensor analysis.) But under *linear* coordinate transformations the ps are constant, and then the sum of tensors even at different points is a tensor. Analogous remarks apply to the product of tensors to be defined next.

If $A \cdots$ and $B \cdots$ are tensors of arbitrary valences, the juxtaposition of their components defines their *outer product*. Thus, for example,

$$C_{klm}^{ij} = A_k^i B_{lm}^j$$

is a tensor of the valence indicated by its indices. (The simple proof is left to the reader.) As a particular case, $A \cdots$ could be a scalar. In conjunction with sum, therefore, we see that any linear combination of tensors of equal valence is a tensor. The outer product of two vectors $\mathbf{A} (= A^i)$ and $\mathbf{B} (= B^i)$ is sometimes written $\mathbf{A} \otimes \mathbf{B}$.

Contraction of a tensor of valence (s, t) consists in the replacement of one superscript and one subscript by a dummy index pair, and results in a tensor of valence $(s - 1, t - 1)$. For example, if A_{klm}^{ij} is a tensor, then

$$B_{km}^j = A_{khm}^{hj}$$

is a tensor of the valence indicated by its indices. (The proof is left to the reader.) Contraction in conjunction with outer product results in an *inner product*, e.g. $C_{ikl} = A_{ij} B_{kl}^j$. A most important particular case of contraction or inner multiplication arises when no free indices remain: the result is an invariant, e.g. A_i^i, A_{ij}^{ij}, $A_{ij} A^{ij}$ are invariants if the As are tensors. (A particular case: $\delta_i^i = N$.)

The last of the algebraic tensor operations is *index permutation*. For example, if the tensor components A_{ij} are exhibited in matrix form, $B_{ij} = A_{ji}$ denotes the components of the transposed matrix, and those components form a tensor, as is immediately obvious from (A.4). Index permutations of all orders are permissible among *either* the subscripts *or* the superscripts of a tensor. Thus we can form such tensor sums as $A_{ij} + A_{ji}$ or $A_{lm}^{ijk} + A_{lm}^{jki} + A_{lm}^{kij}$, and such tensor equations as $A_{ij} = A_{ji}$. It follows, in particular, that the symmetry (or antisymmetry) of a tensor is an invariant property, i.e. is preserved under coordinate transformations.

A9. Differentiation of tensors

We shall write

$$\frac{\partial}{\partial x^r}(A_{\,l\,\ldots\,n}^{\,i\,\ldots\,k}) = A_{\,l\,\ldots\,n,\,r}^{\,i\,\ldots\,k}.$$

(A special case of this notation has already been used in Section A6 in defining $\phi_{,i}$.) Then if $A_{\,l\,\ldots\,n}^{\,i\,\ldots\,k}$ is a (field-)tensor, differentiation of the general tensor component transformation (A.5) yields (by use of $\partial/\partial x^{r'} = \partial/\partial x^r p_{r'}^r$):

$$\mathrm{A}_{\,l'\,\ldots\,n',\,r'}^{\,i'\,\ldots\,k'} = A_{\,l\,\ldots\,n,\,r}^{\,i\,\ldots\,k}\, p_i^{i'} \ldots p_k^{k'}\, p_{l'}^{l} \ldots p_{n'}^{n}\, p_{r'}^{r} + P_1 + P_2 + \ldots,$$

where P_1, P_2, etc., are terms involving derivatives of the ps. [It should be noted that a product with implied summations—like the right-hand side of (A.5)—can be differentiated with complete disregard of these summations, since sum and derivative commute.] Under general coordinate transformations, therefore, $A_{\,l\,\ldots\,n,\,r}^{\,i\,\ldots\,k}$ is not a tensor. But under *linear* coordinate transformations (ps constant) $A_{\,l\,\ldots\,n,\,r}^{\,i\,\ldots\,k}$ behaves as a tensor of the valence indicated by all its indices, including r, since then the Ps vanish. By a repetition of the argument, all higher-order partial derivatives,

$$A_{\,l\,\ldots\,n,\,rs}^{\,i\,\ldots\,k} = \frac{\partial^2}{\partial x^r \partial x^s}(A_{\,l\,\ldots\,n}^{\,i\,\ldots\,k})$$

etc. also behave as tensors under linear transformations, each partial differentiation adding a new covariant index.

Consider a curve in V_N defined by the equations $x^i = x^i(t)$, where t is a scalar (invariant) parameter. Then $\mathrm{d}x^i/\mathrm{d}t$ is a vector under *all* transformations, the proof being similar to that for $\mathrm{d}x^i$. That the scalar derivative of any field tensor, $(d/dt) A^i_{l \cdots n}{}^k$, behaves as a tensor under *linear* transformations follows at once from differentiation of (A.5). We may note how the four basic vectors of classical mechanics—velocity, acceleration, momentum, force—are all built up from the operations of differentiation and multiplying by scalars: dx^i/dt, d^2x^i/dt^2, mdx^i/dt, md^2x^i/dt^2 (t = Newtonian time).

A10. The quotient rule

Although we cannot usefully form a 'quotient' of tensors, an object like C^{ij} in the equation $A^i = C^{ij} B_j$, where A^i and B_i are tensors, can be formally regarded as a kind of quotient of A^i and B_i. This gives the name to a most useful rule for recognizing tensors, the *quotient rule*, which roughly says that the quotient of tensors is itself a tensor. Accurately stated, it reads thus: *If a set of components, when combined by a given type of multiplication with an arbitrary tensor of a given valency yields a tensor, then the set constitutes a tensor.* The general method of proof can be sufficiently indicated by considering the above special case. Suppose we know of the components C^{ij} that for an arbitrary tensor B_i the product $C^{ij} B_j$ is a tensor. Let C^{ij} and $C^{i'j'}$ be the components of our object in two arbitrary coordinate systems $\{x^i\}$ and $\{x^{i'}\}$. Then first by the hypothesis and second by the tensor character of B_i, we have

$$C^{i'j'} B_{j'} = (C^{ij} B_j) p^{i'}_i = C^{ij} B_{j'} p^{j'}_j p^{i'}_i ,$$

whence, for all i',

$$(C^{i'j'} - C^{ij} p^{i'}_i p^{j'}_j) B_{j'} = 0.$$

But $B_{j'}$ in *one* coordinate system is quite arbitrary, so we can successively give it values $(1, 0, 0, \ldots), (0, 1, 0, \ldots)$, etc. and thereby find that the above expression in parentheses vanishes for all j'. This shows that C^{ij} is a tensor.

There are several useful variations of the quotient rule [see, for instance, Exercises A(4) and A(5)]. In particular, if the product involved is an *outer* one, the tensor multiplier need not be *arbitrary*, merely *non-zero*, as should be obvious from the above proof.

A11. The metric

So far no special structure has been assumed for V_N. But many spaces in which tensors play a role are *metric* spaces, i.e., they possess a function ds which assigns distances to pairs of neighbouring points. In particular, one calls a space *Riemannian*[1] if

$$ds^2 = g_{ij}dx^i dx^j, \tag{A.6}$$

where the gs are generally functions of position, and are subject only to the restriction det $(g_{ij}) \neq 0$. They may, without loss of generality, be assumed to be symmetric: $g_{ij} = g_{ji}$. Euclidean N-space, which has $ds^2 = (dx^1)^2 + \ldots + (dx^N)^2$, is only one example. Since we require ds^2 to be an invariant, it follows from a simple variation of the quotient rule [see Exercise A(4)] that g_{ij} must be a tensor. We call it the *metric tensor*, and (A.6) the *metric*.

In metric spaces one often adopts a notation for vectors analogous to that in Euclidean spaces, i.e. one writes **A** for A^i etc. A *scalar product* of two vectors is then defined by

$$\mathbf{A} \cdot \mathbf{B} = g_{ij}A^i B^j. \tag{A.7}$$

This clearly satisfies the relations

$$\mathbf{A} \cdot \mathbf{B} = \mathbf{B} \cdot \mathbf{A}, \quad \mathbf{A} \cdot (\mathbf{B} + \mathbf{C}) = \mathbf{A} \cdot \mathbf{B} + \mathbf{A} \cdot \mathbf{C}$$

and—in the special case of constant gs—the Leibniz rule

$$d(\mathbf{A} \cdot \mathbf{B}) = d\mathbf{A} \cdot \mathbf{B} + \mathbf{A} \cdot d\mathbf{B}.$$

Two vectors are said to be *orthogonal* if their scalar product vanishes. A particular case of scalar product is the *square* of a vector (which in pseudo-Riemannian spaces can be positive *or* negative):

$$\mathbf{A}^2 = \mathbf{A} \cdot \mathbf{A} = g_{ij}A^i A^j. \tag{A.8}$$

From it one defines the (non-negative) *norm* $|\mathbf{A}|$, or simply A, by the equations $A = |\mathbf{A}^2|^{1/2} \geq 0$.

In metric spaces there exists a fifth basic algebraic tensor operation, namely the *raising and lowering of indices*. For this purpose we define g^{ij} as the elements of the inverse of the *matrix* (g_{ij}). Because of the symmetry of (g_{ij}), its inverse (g^{ij}) is also symmetric. The g^{ij} are defined uniquely by the equations

$$g^{ij}g_{jk} = \delta^i_k. \tag{A.9}$$

If $g^{i'j'}$ denote the tensor transforms of g^{ij} in the $x^{i'}$ system [according to (A.3)], then, by the form-invariance of tensor component equations (since g_{ij} and δ^i_j are tensors), we have from (A.9)

$$g^{i'j'}g_{i'k'} = \delta^{i'}_{k'}.$$

But these are also the equations that uniquely define the inverse $(g^{i'j'})$ of the matrix $(g_{i'j'})$. Hence the g^{ij} thus defined in all coordinate systems constitute a contravariant tensor said to be *conjugate* to g_{ij}.

Now the operations of raising and lowering indices consist in forming inner products of a given tensor with g_{ij} or g^{ij}. For example, given a contravariant vector A^i, we define its covariant components A_i by the equations

$$A_i = g_{ij}A^j. \tag{A.10}$$

Conversely, given a covariant vector B_i, we define its contravariant components B^i by the equations

$$B^i = g^{ij}B_j. \tag{A.11}$$

As can easily be verified, these operations are consistent, in that the raising of a lowered index, and vice versa, leads back to the original component. They can of course be extended to raise or lower any or all of the free indices of any given tensor: e.g. if $A_{ij}{}^k$ is a tensor we can define $A^i{}_{jk}$ by the equations

$$A^i{}_{jk} = g^{ir}g_{ks}A_{rj}{}^s.$$

Note how it may sometimes be convenient, for instant recognition, to use dummies from a distant part of the index alphabet. Note also that when we anticipate raising and lowering of indices, we should write the indices in staggered form so that no superscript is directly above a subscript, e.g. $C^{ij}{}_{kl}$ rather than C^{ij}_{kl}. It should also be pointed out that we think of, say, $A_{ij}{}^k$ and $A^i{}_{jk}$ as merely different descriptions of the *same* object.

Lastly, the reader should note and verify

$$\mathbf{A} \cdot \mathbf{B} = g_{ij}A^iB^j = A^iB_i, \tag{A.12}$$

$$g^i{}_j = \delta^i_j, \tag{A.13}$$

the 'see-saw' rule for any dummy index pair:

$$A_iB^i = A^iB_i, \tag{A.14}$$

and the conservation of symmetries:

$$A_{ij} = \pm A_{ji} \Leftrightarrow A^{ij} = \pm A^{ji}. \tag{A.15}$$

One use of index shifting is that it allows us to form inner products of tensors that could not otherwise be so combined (e.g. $A_{ij}B^{ij}$ from A_{ij} and B_{ij}). Another use occurs in the construction of tensor equations, where all terms must balance in their covariant and contravariant indices (e.g. $A_{ij} + B_{ij} = C_{ij}$ from A_{ij}, B^{ij}, and $C^i{}_j$).

It may be mentioned that there is another important use of the metric tensor g_{ij}, namely in the construction of the so-called 'covariant derivative' of tensors under general coordinate transformations. But this is not needed in special relativity (as long as we adhere to 'standard coordinates'), and so will not be dealt with here.

---

[1] Strictly, such a space is called Riemannian only if $ds^2 > 0$ whenever $dx^i \neq 0$, and 'pseudo-Riemannian' otherwise.

Exercises A

1. (i) A vector A^i has components $\dot{x}, \dot{y}$ $(. = d/dt)$ in rectangular Cartesian coordinates; what are its components in polar coordinates? (ii) A vector B^i has components $\ddot{x}, \ddot{y}$ in rectangular Cartesian coordinates; prove, directly from (A.3), that its components in polar coordinates are $\ddot{r} - r\dot{\theta}^2$, $\ddot{\theta} + 2\dot{r}\dot{\theta}/r$.

2. The linear coordinate transformation $x^{i'} = a^{i'}_i x^i$ with inverse $x^i = b^i_{i'} x^{i'}$ is called *orthogonal* if $b^i_{i'} = a^{i'}_i$. Prove that under such transformations there is no difference between covariance and contravariance. (Objects behaving as tensors under orthogonal transformations are called *Cartesian tensors*. In this book we call three-dimensional Cartesian tensors simply 'three-tensors'.)

3. Use the quotient rule to show that δ^i_j is a tensor.

4. If C_{ij} is symmetric and $C_{ij} A^i A^j$ is a scalar for an arbitrary vector A^i, prove that C_{ij} is a tensor. [*Hint*: prove $(C_{i'j'} - C_{ij} p^i_{i'} p^j_{j'}) A^{i'} A^{j'} = 0$, and then choose an $A^{i'}$ with (i) only one non-zero component, (ii) only two non-zero components.]

5. If $C_{ij} A^i A^j$ is a scalar for an arbitrary vector A^i, prove that $C_{ij} + C_{ji}$ is a tensor. (ii) If $C_{ij} A^i B^j$ is a scalar for two arbitrary vectors A^i, B^i, prove that C_{ij} is a tensor.

6. Prove that for any vector A_i the expression $A_{i,j} - A_{j,i}$ is a tensor. Similarly prove that for any antisymmetric tensor E_{ij} the expression $E_{ij,k} + E_{jk,i} + E_{ki,j}$ is a tensor.

7. If the coefficients a_{ij} are constant and symmetric, prove that

$$(a_{ij} A^i A^j)_{,k} = 2a_{ij} A^i A^j{}_{,k}.$$

8. For any object A_{ij} we define the 'symmetric part' $A_{(ij)} = \frac{1}{2}(A_{ij}$

$+A_{ji}$) and the 'antisymmetric part' $A_{[ij]} = \frac{1}{2}(A_{ij} - A_{ji})$. Prove that A_{ij} $= A_{(ij)} + A_{[ij]}$. Analogous definitions are made for B^{ij}. If A_{ij} and B^{ij} are tensors, so are their parts. Prove

$$A_{(ij)}B^{ij} = A_{(ij)}B^{(ij)} = A_{ij}B^{(ij)},$$

and the same identities with square instead of round brackets. Prove also that $A_{(ij)}B^{[ij]} = 0$.

9. If $g_{ij} = 0$ for $i \neq j$, prove that $g^{ij} = 0$ for $i \neq j$, and $g^{ii} = 1/g_{ii}$.

10. Prove that in Euclidean N-space referred to coordinates in terms of which the metric is $ds^2 = (dx^1)^2 + \ldots + (dx^N)^2$, the value of tensor components is unaffected by raising or lowering of indices. [Compare Exercise (2) above; it can be shown that transformations which preserve the Euclidean metric are orthogonal.]

11. Prove that in the case of a constant metric tensor, the operations of partial differentiation and index shifting commute, so that, for example, the relation between $A^i{}_{,j}$ and $A_{i,j}$ is unambiguous.

12. Any curve $x^i = x^i(t)$ in a metric space V_N can be re-parameterized in terms of the arc $s = \int \left| g_{ij} \dot{x}^i \dot{x}^j \right|^{1/2} dt$, so that $x^i = x^i(s)$. Prove that the *tangent vector* dx^i/ds has unit norm, and that the *principal normal vector* d^2x^i/ds^2 (in the case of constant gs) is orthogonal to dx^i/ds. [*Hint*: differentiate the equation resulting from (A.6) on division by ds^2.] *Its* norm measures the arc rate of turning of the unit tangent vector and is accordingly called the curvature of the curve.

13. From Equation (A.2)(ii) deduce the well-known result that the Jacobian matrices $(p_i^{i'})$ and $(p_{i'}^i)$ of the respective transformations $x^i \to x^{i'}$ and $x^{i'} \to x^i$ are inverses of each other.

14. If $\|A_{ij}\|$ and $\|A_{i'j'}\|$ denote the determinants of a tensor A_{ij} in the x^i and $x^{i'}$ systems respectively, prove $\|A_{i'j'}\| = \|A_{ij}\| p^{-2}$, where p $= \|p_i^{i'}\|$ is the Jacobian of the transformation $x^i \to x^{i'}$. Deduce that the non-singularity of A_{ij} is invariant under non-singular coordinate transformations. [*Hint*: $\|a_j^i b_k^j\| = \|a_j^i\| \|b_j^i\|$.]

15. In determinant theory it is shown that, if a_p^i denotes the pth element in the ith row of a 4×4 determinant whose value is a, then $e_{ijkl} a_p^i a_q^j a_r^k a_s^l = e_{pqrs} a$, where e_{ijkl} are the permutation symbols whose value is $+1$, -1, or 0 according as i, j, k, l is an even permutation, an odd permutation, or no permutation at all, of the numbers 1, 2, 3, 4. Deduce that in four-dimensional space the components $\varepsilon_{ijkl} = |g|^{1/2} e_{ijkl}$, where g is the determinant of the metric tensor, behave as a covariant tensor—the *permutation tensor*— under

any group of transformations with positive Jacobians. (The transformations of the Lorentz and Poincaré groups are cases in point: their Jacobians are unity; and the relevant g is -1.) Permutation tensors can, of course, be constructed similarly in spaces of any dimension. [*Hint*: refer to the preceding exercise.]

16. For the permutation tensor defined in the above exercise, prove

$$\text{(i)} \quad \varepsilon^{ijkl} = e|g|^{-1/2} e^{ijkl},$$

where e is the sign of g, and where $e^{ijkl} = e_{ijkl}$;

$$\text{(ii)} \quad \varepsilon_{ijkl}\varepsilon^{ijrs} = e4\delta_k^{[r}\delta_l^{s]} = e2(\delta_k^r\delta_l^s - \delta_k^s\delta_l^r),$$

[*Hint*: consider the three cases (i): $r = s$ or $k = l$; (ii): not (i) and $(r, s) = (k, l)$ or (l, k); (iii): not (i) or (ii).]

17. For any antisymmetric tensor F_{ij} (in four-dimensional space) we define a *dual tensor* $\overset{*}{F}_{ij} = \frac{1}{2}\varepsilon_{ijkl}F^{kl}$. Prove that the dual of the dual is plus or minus the original tensor: $\frac{1}{2}\varepsilon_{ijkl}\overset{*}{F}{}^{kl} = eF_{ij}$.

18. For the dual tensor defined in the above exericse, prove $\overset{*}{F}_{ij}\overset{*}{F}{}^{ij} = eF_{ij}F^{ij}$.

19. Prove

$$\varepsilon_{ijkl}\varepsilon^{ipqr} = e6\delta_j^{[p}\delta_k^q\delta_l^{r]}$$

$$= e\,(\delta_j^p\delta_k^q\delta_l^r + \delta_j^q\delta_k^r\delta_l^p + \delta_j^r\delta_k^p\delta_l^q - \delta_j^p\delta_k^r\delta_l^q - \delta_j^r\delta_k^q\delta_l^p - \delta_j^q\delta_k^p\delta_l^r).$$

[*Hint*: show that each side vanishes if p, q, r are not distinct; if they *are* distinct, show that each side is $+1$, -1, or 0, according to whether j, k, l is an even permutation, an odd permutation, or no permutation at all, of p, q, r.] Use this to re-derive the second result of Exercise (16) above.

20. Using the result of the preceding exercise, prove that, for any two antisymmetric tensors F_{ij} and G_{ij},

$$\overset{*}{F}_{ip}\overset{*}{G}{}^{iq} = e\left[\frac{1}{2}\delta_p^q(F^{jk}G_{jk}) - F^{jq}G_{jp}\right].$$

INDEX

Page numbers in italics indicate the beginning of an entire section dealing with the subject

a higher f is observed when source moves towards
receptor or receptor towards source, as in sound.

A.

A. detector time going slower, w/ is same,
frequency is greater, according to donor

frequency is 2b/y as fast, as detectors

time is slow

if detector moves away from source, all should = out
as waves are less frequent, time is slower

B.

Source's time is slow
detector sees speed at C

freq increases,
therefore w/ decreases -- Check book clock
must get ...

if light source goes in reverse,
freq decreases
w/ increases

Check on first w/ of sound source

p.91 intro to special rel

Show that according to

My & Einstein,

Einstein Ea... ... taking a huge
long time according to its clock,
but not so long according to its
observation of our ships clock.

Ship ... should take real long,
but even lose time by Earth
clock.

P.194. Theory of Relativity based on
physical reality

K. 41 Time to space ...

p. 44-51
p. 3-5 & thats all

DATE DUE